作者简介

程大利 拥有28年在中国市场涉及商业地产、零售及城市综合体等不同类型项目的投资、开发及资产管理的专业管理经验。

她曾服务的跨国企业和国内企业包括：美国西蒙集团、美国华平投资集团、中国光大控股光大安石、摩根士丹利房地产基金、红星美凯龙集团、新希望集团、东方集团、深国投商用置业等。

2017年01月加入百联集团有限公司至今，程女士担任上海百联资产控股有限公司CEO，上市公司百联股份有限公司董事。

About the author

Ms. Cheng Dali has 28 years of experience in investing, acquiring, developing and managing commercial real estate, retail and mixed-use projects in China.
Ms. Cheng joined Bailian Group in 2017 and she is the CEO of Shanghai Bailian Asset Holding Co., Ltd.
The companies she had served include: Simon Property Group (SPG), Warburg Pincus, Everbright Ashmore Real Estate, Morganstanley, Redstar Macalline, The New Hope Group, Orient Group and SZITIC, etc.

Urban Renewal Recordings of Shanghai Bailian Asset Holding Co., Ltd.

RENAISSANCE
Embracing Classics through the History
Renaissance · Integration · Wisdom · Sharing

Edited by: Cheng Dali
Written by: Cheng Dali, Liu Yi, Li Zaili
Produced by: Shanghai Bailian Asset Holding Co., Ltd.
Planned by: Insight Shanghai
Consultant: Shi Wenqiu

百联资控城市更新实录

唤 醒

穿越历史　邂逅经典

唤醒　融通　智慧　分享

程大利·主编

程大利 刘懿 李再励·执笔

百联资控·出品

Insight见著·策划

施文球·顾问

Introduction

Shanghai has been expanding and renewing since its opening in 1843. Excellent buildings of different ages have left classic marks on Shanghai as a witness of history, a display of times and a unique city icon. They retain the memory of Shanghai, embellish Shanghai and play a role in daily life of Shanghainese for generations.
Shanghai has entered the stage of "Organic Renewal", mainly in commercial and office areas, industrial land, old residential areas, and areas with historical cultural features. Shanghai is paying more attention to improving its quality and vitality in order to realize connotative innovation-driven development. Compared with the previous renewal mode of excessive demolition and reconstruction, the renovation and renewal of in-stock properties aim to satisfy contemporary people with the functional needs for buildings and create space with a sense of belonging while preserving the cultural characteristics of the old buildings.
By presenting the architectural features, historical changes and cultural contents of eight in-stock old buildings renovated by Bailian Group in the central urban area of Shanghai, and interpreting their renovation positioning planning, design highlights and challenges, this book comprehensively describes the experience of Shanghai Bailian Asset Holding Co., Ltd. (hereafter referred to as Bailian Asset Holding) in in-stock renovation projects, and illustrates its philosophy and practice of changing Shanghai with innovation and redefining value with operation. Following the renewal clue from buildings to regions and finally to the whole city, this book objectively describes the in-stock renewal projects and analyzes them in tandem, coupled with comprehensive expressions such as historical old photos, architectural status pictures, and planning and design drawings, aiming to record the practical experience of Bailian Asset Holding in urban renewal. For professionals including urban planners, architects and urban development managers, this book vividly records urban renewal concepts and practical strategies; for the public, this book decodes the cultural stories and history of the old buildings, helping them understand the urban development and cultural memory of Shanghai.

内容简介

自1843年开埠以来,上海就一直处于不断的扩展和更新过程中,不同时代的优秀建筑给城市留下了经典的印记,成为历史的见证、时代风貌的展示和独特的城市名片。它们保存着这个城市的记忆,给予着城市温度,上演着一代又一代上海人的生活。

当前,上海已经进入了城市"有机更新"的阶段,主要聚焦于商业商务区、工业用地、老旧住区和历史文化风貌区四个方面,更注重提升城市的品质和活力,致力于实现城市的内涵式创新发展。较之以前"大拆大建"式的更新模式,现在存量物业的改造更新更追求在保留老旧建筑文化特质的同时,满足当代人对建筑的功能需求,并为人们提供一个有归属感的空间。

本书立足于百联集团位于上海中心城区的八个存量老建筑的修缮改造实例,梳理每一栋老建筑的建筑特色、历史变迁和人文底蕴,解读项目的改造定位规划、设计亮点与挑战,全面展现上海百联资产控股有限公司(以下简称"百联资控")对于存量项目改造的心路历程,展示百联资控"创新改变城市、运营重构价值"的理念和实践。从建筑更新升级到区域更新再到整个城市的更新,本书以此为线索将百联资控的存量更新项目作客观表述、串联剖析,并采用历史老照片、建筑现状图片、规划设计图纸等综合表达,旨在记录百联资控之于城市更新的实践经验。对于城市规划师、建筑设计师和城市开发管理者等专业人士而言,它是一本集城市更新理念和实践策略的生动读本;对于社会大众而言,它是一本记录了城市老建筑的历史人文故事和沉浮命运的"城市解码书",对于了解上海的城市发展和文脉记忆亦有助益。

"建筑是世界的年鉴,当歌曲和传说都缄默的时候,只有它还在说话。"

—— 果戈里(俄)

建筑是一个时代审美与技术的综合体现,城市更新是城市的理想、审美和价值的体现。文化、思想、技术、生活方式、商业生态、城市愿景……太多内涵被包容在城市更新的巨大命题之中,推动着当代人在城市开发治理理念和水平上的变革和提升。

每一座城市都在不断求索创新的过程中形成了各自独有的人文风貌。在城市更新的前沿阵地——上海,也涌现了一批又一批蕴含人文特色、展现时代风貌、符合发展需求的城市更新项目,为这座城市风起云涌的进化历程献上了迭次焕新的繁华。

更新 SHANGHAI

東

SHANGHAI

謀

"Building is the yearbook of the world. When the songs and legends keep silent, the building is still talking."
– Gogol (Russia)

Building is the comprehensive embodiment of aesthetics and technology of an era. Urban renewal is the embodiment of urban ideals, aesthetics and values. Culture, ideology, technology, lifestyle, business ecology, city vision … there are too many connotations contained in the huge theme of urban renewal to drive the transformation and promotion of concept and level of contemporary urban development and governance.

The unique humanity landscape is formed by every city in the process of unswerving exploration and pursuit for innovation. In Shanghai, a frontier of urban renewal, also bears witness to batches of urban renewal projects with humanistic features, era styles, and adaptations to development. This contributes to the vigorous evolution of this metropolis.

序一

让历史建筑"活"在城市更新之中

上海这座城市始终处在城市更新的过程中。城市的每一个部分都在发生变化，但仍然保留了能够将它区别于其他城市的底色和特点，那就是城市的历史记忆，是城市文化和社会发展的动力。13世纪的上海这一带"人烟浩穰，海舶辐辏"，从渔村和渔业港口逐渐演变成沿海巨镇，再从1843年开埠历经变迁直至今天的世界大都市，城市始终充满着无限生机与活力。上海在新一轮城市总体规划中的愿景是"迈向卓越的全球城市"，这需要通过新时代的发展模式才能实现。城市开发从过去的"圈地"、"扩张"，转变为更新与复兴的和谐发展，是城市空间和城市功能更深层次的修复与改善提升。

在寸土寸金的上海市中心，星罗棋布着不同时代、多种类型、各国风格的精美历史建筑，是上海弥足珍贵的文化遗产，也是前人智慧的积淀，更是城市内涵、品质、特色的重要标志。如何在城市更新发展过程中，做好历史建筑的保护传承和有效活化利用，一直是各国不断深入研究探索的世界性课题，也是摆在城市管理者面前一个极为现实的重要工作。

我们欣喜地看到，上海百联资产控股有限公司近些年来，在探索城市更新对历史建筑保护和活化利用方面，进行了许多有益的探索和尝试，在历史建筑保护模式和有效利用机制上，积累了实践的经验，实为难能可贵。

百联资控能够组织力量，在历史考证和现状勘测后，为每一幢历史建筑建档立案，并提出针对性的更新目标和设想方案，在行业中做出了表率，尊重、传承和保护利用历史建筑，不仅继承了历史文化，留住了城市记忆，也为企业厘清家底，让历史建筑"活化"，为企业发展增添更为强劲的后发优势，创建企业文化，对历史负责，也对未来负责。

上海历史建筑的保护利用就是要充分发掘其承载的城市历史和文化记忆，让它们"活"在当下，为城市发展、生产生活持续注入精彩的丰富内涵，而不仅仅是复原一座供人瞻仰的古迹。在尊重历史、尊重原貌的前提下，一些建筑也可以融合新业态重塑功能，用"秀色"替换"锈迹"，为续写城市历史文脉做出应有的贡献。我们看到了百联资控在这方面所做的切实努力。同时，也期待更多企业也能深度参与城市历史老建筑的修缮和复兴，提升城市更新水平，助力上海迈向卓越的全球城市。

—— 中国科学院院士

Preface One
Reviving Historical Buildings in Urban Renewal

Shanghai is always in the process of urban renewal, and every part of this city is changing day by day. However, Shanghai still retains the keynotes and characteristics to distinguish it from other cities, which are historical memories on its own and driving forces for urban culture and social development. As early as the 13th century, Shanghai was densely populated, where seagoing vessels of considerable size converged. Shanghai gradually evolved from a fishing village and fishing port to a giant coastal city. Since it was opened as a commercial port in 1843, Shanghai has emerged as a world-class metropolis in the contemporary era, with full of vitality and infinite vigor. Shanghai's vision in a new round of urban planning lies in "Shanghai Moving towards the Excellent Global City", which calls for development model in a new era. Urban development shifts from the previous "enclosure" and "sprawling" to the harmonious development of renewal and revival, which marks restoration and improvement of urban space and urban function in a profounder manner.

The downtown of Shanghai is dotted with exquisite historical buildings of different eras, types and national styles. They are not only precious cultural heritages of Shanghai and treasure houses of the predecessors' wisdom, but also important epitomes of urban connotation, quality and feature. In the process of urban renewal and development, the approaches to protection, inheritance, effective revival and utilization of historical buildings are always worldwide issues for continuous in-depth research and exploration in various countries, which are also practical and important tasks for city administrators to fulfill.

To our joy, in recent years, Bailian Asset Holding has conducted many useful explorations and attempts in groping for protection, revival and utilization of historical buildings through urban renewal, and has gained practical experience. How commendable it is!

Bailian Asset Holding can mobilize forces to develop archive and file case for each historical building on the basis of historical research and current situation investigation, putting forward pertinent renewal goals and scenarios, setting an example in the industry. Respect, inheritance, protection and utilization of historical buildings not only carry on the historical culture and retain the city memories, but also clarify in-house resources, make the historical buildings "alive", add stronger latecomer advantage to the corporate development, and establish corporate culture from the perspective of Bailian Asset Holding. Bailian Asset Holding shoulders a grand mission for the history and the future.

The protection and utilization of Shanghai-based historical buildings are aimed to give full play to urban history and cultural memories contained in these buildings, make they "alive" in the present, and continue to inject a variety of wonderful connotations into development, production and lifestyle of the city rather than just restore them into historical sites worshipped by the people. Under the premise of respecting history and respecting the original appearance, some buildings can also integrate face-lifting function of the new business format, replace "obsolete glory" with "beautiful scenery", and make due contributions to continuing the historical context of the city. We bear witness to the down-to-earth efforts made by Bailian Asset Holding in this regard. Moreover, more enterprises are also expected to dedicate to renovation and revival of the citywide historical buildings, upgrade the level of urban renewal, and boost Shanghai in pursuit for the goal of "Excellent Global City".

— Zheng Shiling, Academician of the Chinese Academy of Sciences

序二
传承融合与城市共生

  建筑所反映和记录的是一个民族的文化情感和国家的历史。一砖一瓦积淀着灰尘走过年华、历经岁月链接着城市的记忆，记录着城市的生活状态，谱写着城市天幕下激昂交错、震撼人心的交响。

  当作为使用者、受益者的我们，一代一代穿行其间，循着古老建筑的步履足迹，尽情感知触碰那些看似质朴简单却承载着历史文化印记的碎瓦颓垣、神工意匠……仿佛漫过岁月，穿越时空，唤醒了城市的记忆。

  城市更新不是一个崭新的话题，可以说更新改造几乎伴随着城市发展的全过程。

  今天，在中国，城市更新之所以被放到一个如此重要的高度，我想是因为我们的城市发展已经走到了一个崭新的阶段，需要对城市功能重新定位。城市更新的背后，是经济体更新、产业更新、区域结构更新、人群组成更新和个体生活的更新。面对城市土地资源稀缺、可持续发展的要求，如何合理地利用有限的城市资源，将人文资源、生态资源融入城市更新，是对一座城市发展能力和创新能力的重要挑战。

  城市更新不是简单的"修旧如旧"，而必须在传承老建筑旧貌的同时注入新的灵魂和血液。我们努力地从时间与空间变化的过程中去认识和理解传承的意义和价值，这正是传统与现代的融合。围绕"尊重历史文脉、延续城市脉络、创建新老建筑对话、多方共建共享"的理念，保护和传承历史文脉、优化公共环境、盘活存量资产效益。在具体手法上，通过对每一座历史建筑的剖析，引进新的场景和故事线索，赋予其新的功能与城市资源，将新与旧、传统与现代有机结合起来，打造新的物理空间，用多元、创新和融合为老建筑注入新的活力。

  在城市更新的项目中，从十里洋场繁华外滩的百年洋行，到苏州河畔经历战争炮火洗礼的仓库，从旧上海金融街的银行旧址到北外滩的"远东第一库"，它们外形风格各有特色，在不同的时期绽放异彩，精巧的装饰细节讲述着不同年代的建筑风格、空间格局与人文历史，具有极高的历史价值和文化价值。

  正因如此，面对这些改造项目时，更需要我们拿出十二分的情怀和敬意，拿出十二分专业度和使命感！从项目前期的历史资料研究、现场勘探考察到方案的修订成型；从历史保护专家（历保专家）评审直至最终施工改造，整个漫长的过程，不容我们一丝懈怠、一丝随意。有的项目，直到进场施工后仍然会发现还有许多之前没有预料到的问题需要紧急抢修，团队也在磨炼中经受了重重考验。

"新生于旧",并不是一件容易的事。老建筑的历史画卷次第展开,让时间和空间续写它的源远流长。但我们深知,在改造中不能割裂老建筑的历史文脉,为它留下可以传承续写的笔触,守护好老建筑方寸之间繁华落尽后的基因密码,使其生命得以延续,是我们肩负的使命。我们重塑的不仅仅是一栋建筑的生命,更是未来城市发展的经络、肌理和脉搏。"江山代有人才出,各领风骚数百年。"老建筑已经矗立了近百年,而我们正在用智慧为她们书写又一个百年的开始。

"建筑是可以阅读的,街区是适合漫步的,城市始终是有温度的。"这种使命感激励我们为造就创新之城、人文之城和生态之城谱写新的篇章!

—— 百联资控CEO

Preface Two
Inheritance, Integration and Urban Symbiosis on the Horizon

Cultural sentiment and the history of a nation are mirrored and recorded by architecture. Bricks and tiles embrace the legacies throughout the ages, cherish important urban memories over time, record living conditions of a city, and compose a resounding and thrilling symphony under the canopy of the heavens.

As users and beneficiaries, we are nestled by architecture for generations. By following the traces of these time-honored buildings, we are immersed into the touch of seemingly simple, plain crumbling walls, dilapidated houses and ingenious designs engraved with the historical and cultural imprints… Going beyond the elapse of time, the memory of a city is awakened through the veil of time and space.

Urban renewal is not a new topic today. Renewal and renovation are almost accompanied by the whole life cycle of urban development.

In today's China, urban renewal has been prioritized because urban development has reached a new level, requiring repositioning of urban functions. Underpinning urban renewal are economic renewal, industrial renewal, regional structure renewal, population composition renewal and individual lifestyle renewal. Faced with the scarcity and sustainable development requirement for urban land resources, the approaches to rationally using limited urban resources and earmarking humanity resources and ecological resources for urban renewal pose severe challenges to the development and innovation capabilities of a city.

Urban renewal is by no means to "restore the old as before"; instead, new essence and blood must be injected into old buildings while carrying forward their historical features. We strive to recognize and understand the significance and value of inheritance in the process of changes in time and space, which merges tradition and modernity in true sense. Upholding the philosophy of "respecting the historical context, continuing the urban vein, establishing dialogues between old and new buildings, and allowing all parties concerned to build and share together", we protect and pass on the historical context, optimize the public environment, and revitalize the in-stock assets to generate benefits. In specific practices, we analyze historical buildings one by one, introduce new scenarios and story clues, endow with new functions and urban resources, as well as organically combine new and old styles and traditional and modern patterns to create new physical spaces for infusing new vitality into old buildings through diversification, innovation and integration.

Our urban renewal projects include not only the century-old foreign firms in the prosperous Bund and the warehouses surviving under the baptism of wars beside Suzhou River, but also the former sites of banks on Financial Street during the Republic of China and the "No.1 Warehouse in the Far East" in the North Bund. Their unique appearances and styles were astonishing in different periods. Exquisite decoration details epitomize architectural styles, spatial pattern and human history of different ages, and indicate extremely high historical and cultural values.

For this reason, we must handle these renovation projects with all sincerity, thorough respect, top-notch professionalism and a sense of mission! The entire far-flung process runs from the historical data research, the on-site exploration and field survey in the early stage of projects, the revision and finalization of drafts, the reviews by historical site protection experts to the final construction and renovation. We must refrain from indolence and inadvertence. In some projects, many unforeseen problems are still exposed and require urgent troubleshooting after we start construction in sites. Our team has also gone through the hoops.

"Reconstructing old buildings into new ones" is a daunting task. Historical pictures of old buildings are unfolded, allowing time and space to continue their history of long-standing. But we are well-aware that the historical context of old buildings should not be separated in the process of renovation. Instead, we should leave them with the brushstrokes that can sustain, safeguard genetic code after the golden years of them, and extend their life as our lofty mission. Remodeling old buildings revives not only their vitality, but also the meridian, texture and pulse of future urban development. "Each era yields a few far better than their peers to take the lead for long." The old buildings have stood for nearly a century, and we pool our wisdom to help them usher in the subsequent century.

"Buildings can be read. Streets are perfect for rambling. Cities are always considerate." This sense of mission inspires us to write a new chapter for building Shanghai into City of Innovation, City of Humanity and City of Ecology!

— Cheng Dali, CEO of Shanghai Bailian Asset Holding Co., Ltd.

序三

旧筑续辉穿越历史，今智授慧邂逅经典

建筑，总是满载某个年代的星辉，闪亮在时光的长河里，以其特殊的方式记录故事、存留情感。只是时间会不可避免地使建筑变得老旧，而随着城市化的进程日益加快，成群的历史建筑转眼间被标新立异的高楼大厦所取代，文化传承的链条出现了明显的裂痕。

对于老建筑，像古董一样保护起来并不是最完美的解决方案，重新赋予老建筑新的空间价值才能让其重获新生，使其更加独特、更加当代。上海的一些老建筑，经过多年的岁月尘封，今由百联资控进行详尽的考察、深度的研究，继而修复、改造，重新赋予使命，给我们带来感动，这无疑是一项饱含情怀和智慧的善举，在上海的城市更新史册上会留下值得记忆的一笔。

旧筑续辉穿越历史，今智授慧邂逅经典。

在历时两年的改造中，百联资控针对不同的老建筑进行了个性鲜明的定位：富有文化底蕴的旧建筑用业态规划来传承城市文脉；富有历史地位的老建筑用品牌创立来彰显城市精神；富有亮丽形态的老建筑用形象重塑来强化城市记忆。

本书所记载的百联资控历史建筑改造项目，将取舍尺度把握得颇有分寸。有的项目不仅仅保存了原有建筑的轮廓风貌，对外部立面的修复尊重历史、古今交融，同时还留下了建筑内部那些富有传承意义、观赏价值、启迪作用的局部和细部结构，再将现代建筑元素引入空间中，这无疑增加了使用功能的内涵。对历史建筑内部空间的设计注重品质，注重协调；对功能业态规划着重需求、恰到好处，均以新时代的匠心精神和精纯技艺来完成，可称得上是城市更新过程中，关于建筑文脉、建筑生命、建筑使命的"三重奏"。

当历史的尘埃拂去，老建筑背后的人文故事亦被重新抽丝剥茧般梳理出来，穿越时空讲述给当代人聆听，延续了建筑的文脉记忆；老建筑的外观、结构被保留、修复、加固，历经百年风霜而再次融入城市景观，延续了建筑的生命价值和城市资产；老建筑的内部空间通过设计师的巧思与时代接轨，被赋予新的功能使命，空间价值被重塑和释放，珍贵的城市土地资源因此而得到最优的运营效益。

凡此种种，都值得其他城市在城市更新与建筑装饰中借鉴、参考，也让本书可以成为涉及城市规划和建筑设计学科的院校的辅助读物。

略谈感想，以此为序。

—— 中国民族建筑研究会专家　资深著作人

Preface Three
Old Buildings Span over the History with Sustained Magnificence, and Contemporary Wisdoms Add Luster to the Classics with Well-established Practices

Architecture always epitomizes an era, recording stories and retaining emotions in its unique way in the long history. However, the time inevitably scars all buildings with bygone traces. With the acceleration of urbanization, groups of historical buildings have been replaced by modern high-rises in a blink of an eye. As a result, the culture succession chain is evidently fractured to some extent.

Instead of protecting like antiques, the optimal solution for old buildings is to give them a second life with new spatial values, so that they can be more unique and contemporary. Some old buildings in Shanghai preserved for years have been repaired, renovated, and endowed with a new mission by Bailian Asset Holding after detailed survey and in-depth research. What an impressive action! This is undoubtedly a philanthropic and wise good deed, which will unforgettably go down in the history of Shanghai's urban renewal.

Old buildings span over the history with sustained magnificence, and contemporary wisdoms add luster to the classics with well-established practices.

In this two-year renovation campaign, the old buildings have been distinctly positioned by Bailian Asset Holding: planning of business forms for old buildings with profound cultural legacies to pass on urban context; brand creation for the old buildings with high historical status to highlight urban spirit; image redefinition for old buildings with bright appearance to reinforce urban memory.

In the old building reconstruction projects by Bailian Asset Holding recorded in this book, the degree of trade-off is properly grasped. Some projects not preserve the original features of old buildings, but also retain partial and subtle structures with inheritance significance, ornamental value and inspiring effects inside the buildings together with modern architectural elements, which undoubtedly beef up the connotation of usage functions. Bailian Asset Holding values quality and coordination of the interior design of old buildings as well as echoes with the demand for the functional format planning of old buildings to the right point. All of them have been completed with the spirit of ingenuity and sophisticated skills in the new era. They mirror the architectural context, architectural life and architectural mission in the process of urban renewal.

When the dust of history blows away, the cultural stories underpinning the old buildings are sorted out again to continue their cultural memories. With appearances and structures preserved, repaired, and reinforced, old buildings have been integrated into urban landscape again to continue their life value and urban assets. Through the designers' ingenious thinking and integration with the times, the internal space of old buildings has been given a new functional mission. Their space value is remolded and released, making the precious urban land resources get the optimal operation benefit.

These examples can be used for reference in urban renewal and architectural decoration by other cities, and this book can also become an auxiliary reading for disciplines of urban planning and architectural design in universities and colleges.

My humble opinions are briefed as above to contribute to this preface.

— Shi Wenqiu, Expert of National Architecture Institute of China and Senior Writer

前 言

建筑重生，唤醒城市新价值

　　1916年，美国记者玛丽·宁德·盖姆韦尔（Mary Ninde Gamewell）在其出版的《通向中国的门户：上海的景象》（The Gateway to China: Pictures of Shanghai）一书中写到："整座城市都处在持续不断的变化中，日复一日，旧建筑正在消失，取而代之的是更现代的建筑。人们担心许多古老的地标很快就会不复存在。"这一描述似乎与100年来上海的变化十分贴切，甚至可以说中国许多城市都面临着这样的窘境。尤其是历经20世纪80年代至今的大规模城市建设，由于人们忽视历史建筑和城市文脉的保护，相当一部分经典老建筑被推倒拆毁，或被改建得面目全非，许多凝聚几代人美好记忆的地方也永远消失了，城市中沉淀的历史魅力也在一天天地丧失。

　　城市的文化和记忆需要得到保存和传承，而这些历史文化记忆的痕迹，恰恰可以在老建筑里找到。对老建筑的保护和更新、活化和利用，已成为城市更新的重要途径之一。

我国著名建筑学专家、中国科学院院士、同济大学教授郑时龄先生在《上海的城市更新与历史建筑保护》一文中指出："城市更新是动态的,既涉及物质性的更新,也涉及非物质性的更新,包括城市结构和空间的更新、建筑的更新、城市环境和道路的更新,更重要的是思想和生活方式、城市治理模式的更新。"经历了大刀阔斧的快速建设阶段之后,上海对建筑文化遗产的保护也越来越趋于理性和成熟,将"更新"建立在充分保护原有历史建筑遗产的基础上。郑时龄院士亦指出,"不同于其他文物,历史建筑不能只是作为博物馆加以封存,绝大部分历史建筑都应当在使用中保护,但是也需要尊重建筑的真实环境和历史风貌。"

上海百联资产控股有限公司对于老建筑的改造更新向来秉承"尊重历史、继承精华"的态度,探索符合上海实际的保护模式,既要延续老建筑的历史特色和文脉记忆,又要赋予其适合时代和未来发展的功能。百联资控注重项目从历史调研、定位策划、留改甄别、规划设计到招商运营全过程精细化管理,在"破"与"立"的平衡之间,充分整合百联集团的商贸资源,为老建筑赋予新的灵魂和使命。

本书记录了百联资控在老建筑更新和重塑方面展开实践的项目。更新空间,是为了提升硬件设施,更好地适应现代生活需求;重塑功能,是为了使之有机活化、焕发新的价值。更新与重塑的过程中,既有团队对历史的思考,也有对未来的憧憬。这种继往开来的改造,为老建筑注入了新的生存意识,真正赐予了建筑的重生!

Foreword
Rebirth of Buildings Awakens New Values of a City

In 1916, American journalist Mary Ninde Gamewell wrote in her book *The Gateway to China: Pictures of Shanghai*: "The whole city is amid constant changes, day after day, old buildings are disappearing and being replaced by more modern buildings. There are fears that many ancient landmarks will soon disappear." This description seemingly quite fits the changes in Shanghai over the past 100 years, and many cities in China are facing such a dilemma. Especially, after the large-scale urban construction since the 1980s, due to the neglect of protection of historical buildings and urban context, a fairly large number of classic old buildings have been demolished or rebuilt beyond all recognition, many places accommodating good memories of generations have vanished, and the historical charm of the city is disappearing.

Urban culture and memory need to be preserved and inherited, whose traces can be found in the old buildings. The protection, renewal, activation and utilization of old buildings have become one of the important means of urban renewal.

Zheng Shiling, famous architectural expert, academician of Chinese Academy of Sciences and professor of Tongji University, pointed out in his article *Urban Renewal and Protection of Historical Buildings in Shanghai* that "Urban renewal is dynamic, involving both material and immaterial renewal, including the renewal of urban structure and space, the renewal of buildings, the renewal of urban environment and roads, and more importantly, the renewal of ideology, lifestyle and urban governance mode." Following the rapid massive construction, Shanghai has been more rational and mature in the protection of architectural cultural heritage: basing renewal on the full protection of the original historical architectural heritage. Zheng Shiling also argued that "Unlike other cultural relics, historical buildings can not only be stored as museums, most of which should be protected in use, whose real environment and historical features also need to be respected."

As for the renovation and renewal of old buildings, holding the attitude of "respecting their history and inheriting their quintessence", Bailian Asset Holding has been exploring a protection mode that accords with the actual situation in Shanghai to extend the historical characteristics and context memory of these old buildings, as well as give them the functions suitable for the development of the times and future. Great importance has been attached to the fine management of these projects in the whole process involving historical survey, positioning planning, retention and renovation screening, design, investment promotion, and operation. Bailian Asset Holding fully integrates its business resources by balancing between demolition and construction, and endows the old buildings with new soul and missions.

This book records the renewal and remodeling projects of old buildings carried out by Bailian Asset Holding. Renewing space is to enhance the hardware facilities for better adaption to the needs in modern life, while remodeling functions is to make them be organically activated and produce new values. In the process of renewal and remodeling, Bailian Asset Holding team has thought about the history and looked forward to the future as well. In this way, renovation injects new awareness of survival into old buildings, and helps them be reborn in true sense!

前言
FOREWORD

建筑的美好，在于它总是沉默着，
在门窗墙壁的每一条缝隙中渗透出生命的力量，
刻画着历史，时间愈久愈显沉寂大气。

The beauty of building lies in its eternal silence. Cracks of the walls, doors and windows are revealing the power of life and depicting the history. The longer time is, the more silence they would generate.

CONTENTS | 目录

序一	让历史建筑"活"在城市更新之中	
Preface One	Reviving Historical Buildings in Urban Renewal	06
序二	传承融合与城市共生	
Preface Two	Inheritance, Integration and Urban Symbiosis on the Horizon	08
序三	旧筑续辉穿越历史，今智授慧邂逅经典	
Preface Three	Old Buildings Span over the History with Sustained Magnificence, and Contemporary Wisdoms Add Luster to the Classics with Well-established Practices	12
前言	建筑重生，唤醒城市新价值	
Foreword	Rebirth of Buildings Awakens New Values of a City	14

第一部分　历史保护建筑集群—传承经典赋能价值
PART 1　Protected Historical Building Clusters : Inheriting Classics and Enabling Values　25

访谈：老建筑之于未来，文化与匠心
Interview: Old Buildings for the Future, Culture and Ingenuity　28

第一章　传奇再生：外滩卜内门洋行（四川中路133号）
——从历史走向未来，经典绽放城市

Chapter One　Legendary Rebirth: Brunner Mond & Co. in the Bund (No. 133 Middle Sichuan Road)
–From History to Future, Classics Stand Forever in Shanghai　32

岁月刻画，前世今生
From the Past to the Present　39

旧日洋行的百年变迁
Centurial Changes in Brunner Mond & Co. Building　39

阅尽千帆的"历史中人"
A Witness Living near the Brunner Mond & Co. Building　44

世纪轮回中的砖瓦诉"情"
Lofty Sentiments Hidden in Bricks and Tiles　48

精致典雅的历史笔触 Exquisite Decorations	50

好"骨架"与新"灵魂"
Superb Framework and New Soul
57

传承历史与共融共生 Inheritance of the History and Integration into the Modern Time	57
新与旧的活力交汇 Dynamic Convergence between New and Old Elements	62

经典邂逅与价值再生
Encounter with Classics and Rebirth of Value
64

时代的映射：墙面雕塑 Mirror of the Time: Wall Sculptures	64
世纪之门：券门 Gate of the Century: Arch Gate	70
锁住老故事：窗 Sealing the Old Story: Windows	72
时间的背面：首层进厅及门厅 Reverse Side of Time: Foyer and Lobby on the Ground Floor	74
找回昔日旧貌：一层夹层 Back to the Past: Mezzanine on the Ground Floor	78
空间的诗学：总经理室 Poetics of Space: General Manager's Office	80
优雅的驻足：各楼层电梯厅 Graceful Elevator Hall on Each Floor	86

项目手记
Project Notes
88

第二章　产业复兴：苏河湾文创艺术体验空间（光复路127号）
——以新业态为导向的城市更新

Chapter Two Industrial Revitalization: Suzhou Creek Cultural and Creative Art Experience Space
(No. 127 Guangfu Road) – Urban Renewal Oriented by New Business Forms ... 90

溯源：时代变迁的见证
Sourcing: Witness to the Changes of the Times ... 94

循迹：于空间中触摸建筑的温度
Tracking: Touch the Temperature of the Building in the Space ... 100

更迭：为建筑遗产吹入生命的气息
Changing: Inject Life into Heritage Buildings ... 106

鼎新：在新旧平衡中找到独特建筑语言
Innovating: Find a Unique Building Language in the Balance between New and Old Elements ... 110

项目手记
Project Notes ... 119

第三章　创新驱动：百空间（天津路50-66号）
——身在"故事"、创造"故事"

Chapter Three Innovation Driving: 百空间 (No. 50-66 Tianjin Road)
—Create a New "Story" in the Old "Story" ... 120

演绎旧上海金融往事
Recurrence of the Financial Story in Old Shanghai ... 126

沿袭书写历史文脉
Inheritance of Historical Context ... 132

创新激发城市活力
Stimulation of Urban Vitality with Innovation ... 136

项目手记
Project Notes ... 147

第四章　经典重启：凯恩宾馆（香港路117号）
——新的空间维度，解构建筑生命

Chapter Four Recurrence of Classics: Kaien Hotel (No. 117 Xianggang Road)
—Deconstruct the Life of Building at a New Spatial Dimension ... 148

采撷不尽的旧日繁花
Countless Old Stories ... 152

邂逅昔日的古典雅致
Encounter with the Classical Elegance in the Past ... 156

新中存旧的跨时空交谈
Talk between the Old and the New across Time ... 160

项目手记
Project Notes ... 166

第五章　文化融合：贵州路263号（原陈英士纪念堂）
——从老建筑的新生看历史街区的更新

Chapter Five Cultural Integration: No. 263 Guizhou Road (Former Chen Yingshi Memorial Hall)
—Renewal of Historical Blocks from the Perspective of the Rebirth of Old Buildings ... 168

走进百年历史的底蕴
Tap into the Essence of the Centurial History ... 170

倾听旧时光的衷肠
Listen to the Touching Story in the Past ... 174

打造"海派硅巷"
Create a " Silicon Alley in Shanghai " ... 180

项目手记
Project Notes ... 184

第二部分 北外滩金融产业办公集群——引领区域产业升级
PART2　North Bund Financial Industry Office Cluster : Lead the Regional Industrial Upgrading　　187

第六章　梦想起航：永兴仓库（杨树浦路61号）
　　　　　——发挥时代功能的空间再造
Chapter Six　Sail the Dream: Yung Shing Warehouse (No. 61 Yangshupu Road)
—Function the Space Reconstruction of the Times　　192

曾经的"远东之最"
Past Fame in the Far East　　200

气势宏大的建筑之韵
Magnificent Construction Style　　205

躺进旧时光里的重生
Rebirth in the Old Days　　209

项目手记
Project Notes　　215

第七章　未来绽放：马登仓库（杨树浦路147号）
　　　　　——基于场所精神的老建筑改造
Chapter Seven　A Bright Future: Pier Marden (No. 147 Yangshupu Road)
—Renovation of the Old Building Based on the Spirit of Place　　216

类叠历史——过去未去
Overlap on the History : The Past Has Not Gone　　218

忠于文脉——传承重构
Loyalty to the Context : Inheritance and Reconstruction　　226

创新空间——未来已来
Innovation Space : The Future Has Come　　233

项目手记
Project Notes　　234

第三部分　莘荟 社区商业产品线——融合创新的活力生活舞台
PART3　莘荟 Community Commercial Project Line : A Dynamic Life Stage of Integrated Innovation　　237

第八章　活力舞台：北宝兴路624号
——立足于立体街区形态的全生活空间

Chapter Eight　A Dynamic Stage: No. 624 North Baoxing Road
—A Dynamic Community Based on Three-dimensional Block Form　　240

老仓库的见证
Witness of the Old Warehouse　　242

活力空间
Dynamic Space　　247

创新即新
Innovative Renovation　　250

品质生活
Quality Life　　258

项目手记
Project Notes　　260

后记
Postscript　　262

参考文献
References　　267

第一部分 PART 1

历史保护建筑集群——传承经典赋能价值
【寻找失落的城市"遗传密码"】

Protected Historical Building Clusters: Inheriting Classics and Enabling Values
Looking for the Lost "Genetic Code" of a City

唤醒
RENAISSANCE

"建筑是成长变化的，

就如同生物组织：

他们由携带和记录遗传密码信息的基因细胞构成。

每一个新的历史阶段，

建筑都在不断地接纳新的信息，

并将其作为新的城市符号。"

——百联资控CEO 程大利

2019年，百联资控遵循黄浦区规划中对于外滩历史风貌保护区持续推进保护性开发的原则，着眼于传承历史，立足长远，推进卜门内洋行、四行仓库（光三分库）保护性修缮，赋予历史保护建筑新的生命，让老建筑焕发新的活力和光彩。

产业建筑携带有反映产业特色和人文活动的历史信息，百联资控运用"保护外貌，改善内部，提升功能"的方法，保持设计初衷，遵循周边文脉，把历史保护老建筑的元素和特征组合到新的使用空间中，不仅在建筑材料、建筑色彩、建筑风格等方面充分考虑与周围历史风貌的协调，体现历史建筑风貌完整性，而且能够把历史建筑的保护利用和城市规划、经济发展、文化旅游、艺术创意活动等结合起来，最大限度地体现人文空间的回归，让历史建筑文化遗产融入现代生活，赋予其新的功能和业态，使其对公众发挥更大价值，从而成为优秀的文化传播的基地和源泉，更让生活工作在这里的人们获得精神的归属感和参与的责任感。

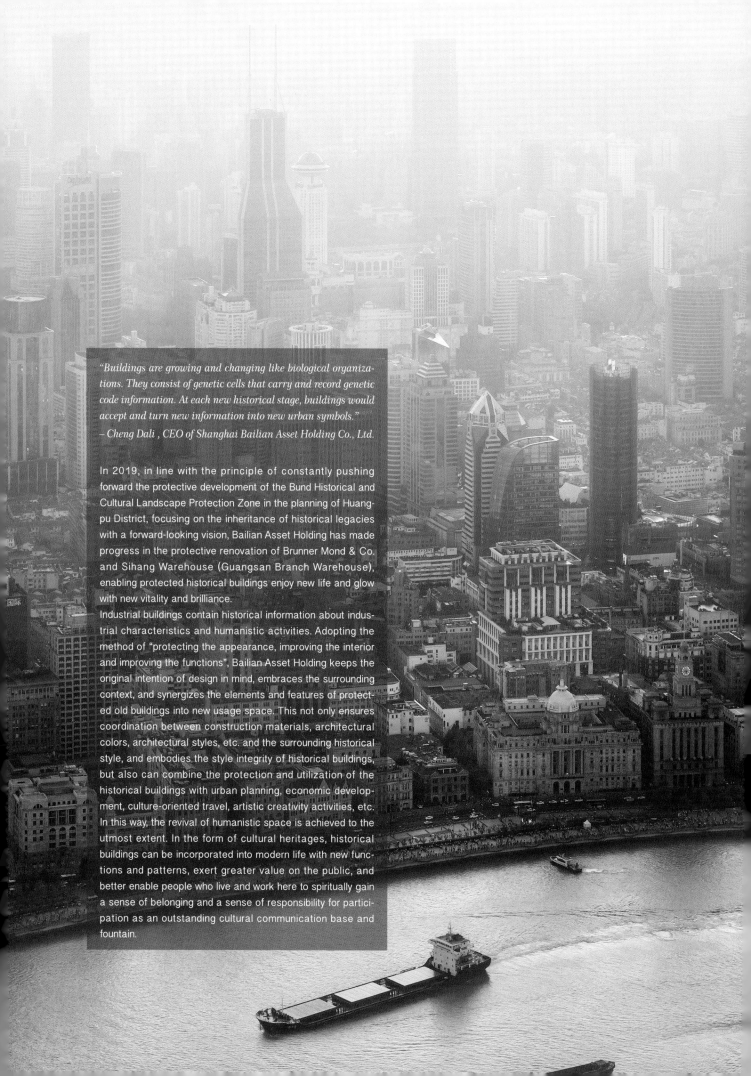

"Buildings are growing and changing like biological organizations. They consist of genetic cells that carry and record genetic code information. At each new historical stage, buildings would accept and turn new information into new urban symbols."
– Cheng Dali, CEO of Shanghai Bailian Asset Holding Co., Ltd.

In 2019, in line with the principle of constantly pushing forward the protective development of the Bund Historical and Cultural Landscape Protection Zone in the planning of Huangpu District, focusing on the inheritance of historical legacies with a forward-looking vision, Bailian Asset Holding has made progress in the protective renovation of Brunner Mond & Co. and Sihang Warehouse (Guangsan Branch Warehouse), enabling protected historical buildings enjoy new life and glow with new vitality and brilliance.

Industrial buildings contain historical information about industrial characteristics and humanistic activities. Adopting the method of "protecting the appearance, improving the interior and improving the functions", Bailian Asset Holding keeps the original intention of design in mind, embraces the surrounding context, and synergizes the elements and features of protected old buildings into new usage space. This not only ensures coordination between construction materials, architectural colors, architectural styles, etc. and the surrounding historical style, and embodies the style integrity of historical buildings, but also can combine the protection and utilization of the historical buildings with urban planning, economic development, culture-oriented travel, artistic creativity activities, etc. In this way, the revival of humanistic space is achieved to the utmost extent. In the form of cultural heritages, historical buildings can be incorporated into modern life with new functions and patterns, exert greater value on the public, and better enable people who live and work here to spiritually gain a sense of belonging and a sense of responsibility for participation as an outstanding cultural communication base and fountain.

唤醒 RENAISSANCE

访谈：老建筑之于未来，文化与匠心

Q：老建筑改造是提升城市形象的紧迫需要，是促进发展的有力措施，那么，您认为历保建筑的建设修缮应该坚持什么样的原则呢？

A：旧城改造不仅仅能满足提升城市形象的需求，随着中国城市化进展和产业发展，必须改造不合理的旧城区来满足全新的需求，包括人口增长和产业发展的需求。但是在城市改造的过程中，我们不能一味否定过去，推倒所有的老建筑来造新房子。有的建筑是历史保护建筑，那你一定不能动。但是有些旧厂房，不是历史保护建筑，可以推倒重建，也可以保留加以改造修缮，这个时候需要在设计前做全面的分析定位，既要对周围环境进行分析，也要对既有建筑进行分析，同时还要对将来的用户进行分析定位。这个原则完全是建立在调查分析的基础之上，目的是满足最终用户的需要。

Q：改造或修缮总是涉及旧建筑选择性保留的问题，但是选择的标准是什么？对于过去的建筑，我们又如何去判断它的好和差？

A：哪种建筑才值得保留是有一定标准的。政府有一个历史文化保护建筑名单，名单上的建筑是不能被拆除的。但有很多不在此列的建筑同样具有很高的价值。我们需要认真鉴定它们的价值，要考虑场地、周围环境、建筑结构、建造方法、材料和很多其他因素。

Q：像卜内门洋行这样的老建筑，地处外滩历史文化风貌区，不管是从建筑的价值还是历史的意义来看，都是很宝贵的资产。对这样的老建筑进行改造或修缮，百联资控会尊崇哪些原则，或者说有哪些心得或经验是可以放在这个项目上的？

A：对于这个项目的改造，可以说远不仅限于用"哪些"心得或经验，而是倾"全部"来做。

第一点，我们要焕发老建筑的历史文物价值和文化价值。任何东西如果没有文化背景，没有历史底蕴，把它做得再漂亮，也只是一个"壳"。把它的历史文化价值挖掘出来，这是第一位的。

第二点，我们把它拿过来之后，要慎重定位今后干什么。过去它可能是洋行，那未来要做什么就是我们的商业运营价值所在。

第三点，才是我们最后的整体效益的问题。因为改造和修缮完成后，它不是一尊佛供在那里，供着就没有意义，要思考如何和我们现在的社会发展结合起来，能发挥什么作用。

最后也是最重要的一点，不管是老建筑还是新建筑，我们做完之后要有一个新老平衡。因为新建筑主要是为了经济效益，或者说是商业价值。但是老建筑要考虑其历史性、文化性和未来性。它已经存在了100年，让它在未来的多少个100年内还可以继续存在，需要我们真的要有些情怀，牺牲其部分商业价值来重塑它的文化价值。

Q：在保持现有文化传承的前提下，如何加入一些符合时代的、现代的元素，或者说如何保持这两者之间的平衡？

A：对一个项目进行改造或修缮设计的时候，要尊重老建筑的原有风格，新的设计方案要和规划相互协调。在此过程中，我们要展现出新老建筑的对比，保持原有建筑的特色并增加具有现代风格的新结构，最后呈现出来的是新老建筑充满活力的交汇。我们坚信这种新旧交融能为用户带来耳目一新的感受。

Q：现在有很多历史建筑改造后成为饭店或者一些

商业用途的项目。请问您是如何看待历史建筑与商业用途相结合这一问题的？

A：历史建筑的新功能不是拍脑袋想出来的，这个和我之前提到的原则一样，需要经过仔细的调查分析，结合周边环境和用户群体得出结论，这也是最终项目能否成功的指标之一。办公、商业、酒店、文艺空间、甚至公寓等都可以作为建筑改造修缮后的新功能，目前就有相当一部分老建筑成为商业建筑。一方面，这与项目的地理位置有关，许多老建筑地处城市中心区域，比较适合发展商业。另一方面，老建筑的文化历史内涵也容易吸引消费者，可以通过这种途径进行文化传播交流。我个人很认同历史建筑和商业相结合，既保护了珍贵的老建筑，又带来商业利益，发挥其应有的城市价值。

Q：现在对老建筑的改造是第一步，后续在运营管理方面，百联资控会整合集团内哪些资源使老建筑真正地"活"起来？

A：从刚开始的策划方案到可研方案，再到施工和运营，都由我们的专业团队操作。我们在刚开始启动时就想好了我们要干什么，我们是从后往前想。在做的过程中当然会结合我们百联的内部资源，贯穿于始终，这些都将成为我们改造、运营的一部分。不仅仅是建筑设计，还有人文环境，还有带来的消费习惯，以及使用者将来对它的适应性，都要达到一个平衡。

Q：近年来，中国建筑改造兴起一股"仿古潮"。可是真正能够改造或者保护得尽如人意的并不是很多，许多改造其实只是将现代建筑和生活模式披上复古的外衣，所生成的建筑或者片区可以说是让人啼笑皆非。那么您觉得具有历史价值的建筑改造中有什么注意事项？

A：改造修缮老建筑是为了保留一部分建筑精华，并使建筑符合现代需求，是一个取其精粹去其糟粕的过程。第一步也是最重要的就是了解老建筑。每个建筑建于不同时期，用途也不尽相同，因此建造方法和使用材料也不一样。为了能够更好地了解老建筑，建筑师必须要做很多调查研究和文献查阅工作。第二步，开发团队要考察这些建筑潜在的新功能，并不是每个建筑单体都能随意地被赋予新功能，所以不能忽视原有的建筑结构来做全盘改造，但可以拆除一些结构上多余的建筑元素来布局新功能。第三步才是对以旧结构和新功能为基础的建筑单体进行重新设计。在老建筑改造过程中，要注意细节处理。新型材料的应用应当符合当今时代的要求，也为原有的老建筑提供必要的补充。

程大利女士在四川中路133号现场考察
on-site survey of No. 133 Middle Sichuan Road by Ms. Cheng Dali

Interview: Old Buildings for the Future, Culture and Ingenuity

Q: The renovation of old buildings is an urgent need to improve the image of a city and an effective measure to promote development. Which principle do you think should be followed in the renovation of protected historical buildings?

A: The renovation of old urban areas is more than satisfaction of the needs to improve the image of a city. With the progress of urbanization and industrial development in China, it is necessary to renovate the unreasonable old urban areas to meet up new needs arisen from population growth, industrial development, etc. In the process of urban renewal, however, it is unwise to blindly negate the past and demolish all the old buildings to construct new ones. Some of old buildings are protected historical buildings excluded from reconstruction, while some old factories are not protected historical buildings and can either be dismantled and rebuilt or renovated. A comprehensive analysis and positioning before design is required, including analyzing the surrounding environment and old buildings, and positioning future users. This principle is based entirely on survey and analysis, aiming to meet the needs of final users.

Q: Renovation or repair always involves the selective retention of old buildings, but what are the criteria? And how to judge the quality of old buildings?

A: There are certain criteria for which buildings are worthy of retention. The government usually issues a list of protected historical and cultural buildings excluded from demolition. But there are also many buildings not specified in such a list but at high values. We should carefully identify their values with full considerations given to a great number of factors such as the site, the surrounding environment, the architectural structure, the construction method, and the materials.

Q: Old buildings like Brunner Mond & Co. in the Bund Zone with Historical and Cultural Features are valuable assets in terms of both architectural value and historical significance. To renovate or repair them, which principles are adopted by Bailian Asset Holding or what kind of expertise or professional experience can be applied for these projects?

A: We renovate such old buildings by sparing no effort, instead of relying on merely some expertise or professional experience. First, the historical relic value and cultural value of old buildings should be revitalized. Anything without cultural background or historical profundity would be nothing however it is beautiful. We prioritize excavating the historical and cultural value of old buildings.

Second, positioning should be carefully clarified. For example, determining the positioning of a foreign firm in the past is the value of our commercial operation.

Third, the overall benefit must be generated. After renovation and repair, the old buildings should not be meaninglessly worshipped like Buddha statues. How to play their roles in the context of the current social development should be considered.

Last but not least, the balance between old and new buildings after renovation should be made. New buildings are mainly constructed for economic benefit or business value. But for old buildings, their historical, cultural and future features are to be considered. These century-old buildings are expected to sustain for more centuries. We should do our best to reconstruct their cultural value at the cost of some commercial interests.

Q: On the premise of sustained inheritance of the existing culture, how to add some modern elements abreast of the times or how to balance between the two?

A: In the renovation or repair design of a project, it is necessary to respect its original style and align the new design scheme with the original plan. In this process, we should give a contrast between the old and new buildings, maintain the characteristics of the original building, add new structures in modern style, and finally perform the vibrant symphony of the old and new buildings. In this way, we firmly believe that users can find everything fresh and new.

Q: Many historical buildings have been renovated into hotels or other commercial projects. What do you think about the combination of historical buildings and commercial use?

A: New functions of historical buildings are not imaginary. This is the same as the principle I mentioned above. It is necessary to carry out careful survey and analysis before drawing any conclusions in the context of surrounding environment and the user community, which is also a measure for whether the final project is successful. Office, business, hotel, art space and even apartment can be new functions of old buildings after renovation or repair. Today, quite a few old buildings are for commercial use. On the one hand, this is related to the geographical location of the projects. Many old buildings are in the central areas of a city and are suitable for commercial development. On the other hand, the cultural and historical connotations of old buildings easily attract consumers, allowing the development of cultural communication. Personally, I agree on the combination of historical buildings and commercial use, which has not protected the precious old buildings, but also generated commercial benefits and given full play to the merited value of a city.

Q: Current renovation of old buildings is the first move. In the subsequent operation and management, what kind of internal resources will be utilized by Bailian Asset Holding to make the old buildings truly "come to life"?

A: Our professional team will fulfill all tasks including initial scheme, feasibility study plan, construction and operation of old buildings. We make clear what to do in a result-oriented manner from the very start. In this process, we will certainly give full play to the internal resources and utilize them as part of our renovation and operation. Perfect harmony should be achieved in not only architectural design, but also humanity environment, consumption habit and users' adaptability to old buildings in the future.

Q: In recent years, "antique tide" has swept in the field of building renovation across China. However, few old buildings can be truly renovated or protected satisfactorily. Many of them are just retro-styled in modern buildings and living patterns, delivering ridiculous buildings or blocks. In your opinion, which Do's and Don'ts should be paid attention to in the renovation of old buildings with historical value?

A: Renovation or repair of old buildings aims to preserve some essences to make them echo with modern needs. It is a process of taking the essence and discarding the dregs. The first and most important thing is to understand them. Different buildings were constructed for different purposes in different periods, making their construction methods and materials different. In order to better understand them, architects must conduct extensive survey, research and literature review. Secondly, our development team should examine the potential new functions of these old buildings. Not all buildings can be endowed with a new function at will. Therefore, the original structure cannot be ignored prior to a thorough renovation, however, some structurally redundant elements can be removed to lay out new functions. Thirdly, redesign should be made based on old structures and new functions. In the renovation process of old buildings, attention should be paid to details. The application of new-type materials should not only meet the requirements of the contemporary era, but furnish indispensable supplements to the original old buildings.

唤醒
RENAISSANCE

第一章

传奇再生：外滩卜内门洋行（四川中路133号）

——从历史走向未来，经典绽放城市

Chapter One

Legendary Rebirth : Brunner Mond & Co. in the Bund (No. 133 Middle Sichuan Road)

– *From History to Future, Classics Stand Forever in Shanghai*

卜内门洋行窗景	外立面历史照片
window scenery of Brunner Mond & Co.	historical photo of appearance

　　在外滩上空，每隔十五分钟会响起一次海关大钟奏鸣的声音。

　　在距"老外滩"仅一街之隔的四川中路（近福州路）上，有一座十分气派的古典式建筑，这就是20世纪20年代前后赫赫有名的"洋碱大王"卜内门公司所在地。

　　卜内门洋行位于外滩历史风貌保护区范围内，区域内的老房子风格多样、中西合璧，是目前上海规模较大、保存较完整、建筑形式丰富的历史建筑群。卜内门洋行大楼于1923年落成，平面为一长方形，现为上海市第二批优秀历史性保护建筑。大楼四面东至四川中路、西至江西中路、北至福州路、南至广东路。

　　站在大楼五层的窗口前，看黄浦江江水缓缓流淌，可以从钟鸣的长短不同，听出具体的时刻来。处在这样位置的老建筑，仿佛百年间也刻上了独有的岁月情怀。

Over the Bund, the clock on top of the Customs House chimes every 15 minutes.
On the Middle Sichuan Road (near Fuzhou Road), just one street away from the old Bund, there stands an imposing classical building, where the famous soda leader Brunner Mond & Co. in the 1920s is situated.
Brunner Mond & Co. lies in the Bund Zone with Historical and Cultural Features, the old buildings in which are in diverse styles with mixed Chinese and Western elements. It is the largest, well-preserved historical building complex in rich architectural forms in Shanghai. The building was completed in 1923, with a rectangular plane. It has been listed in the second batch of excellent historic protection buildings in Shanghai. It stretches to Middle Sichuan Road in the east, Middle Jiangxi Road in the west, Fuzhou Road in the north and Guangdong Road in the south.
Standing by the window on the fifth floor of the building, you can see the Huangpu River flowing slowly and tell the specific time from the length of the chiming. In such a position, this century-old building seems to have been engraved with a unique mark of time.

　　外滩区域内保留着一批建于1920—1936年间的各式近代西洋建筑,为外滩历史文化风貌区的核心区域,是外滩"万国建筑博览会"的源头,也是上海现代城市的源头。

　　黄浦区在历史风貌保护上已基本形成由"历史风貌建筑—风貌保护道路—历史文化风貌区"构成的"点、线、面"相结合的复合型历史风貌空间体系。

　　与外滩建筑群相邻的卜内门洋行大厦,是黄浦区保护与提升的标杆性工程,它的改造和更新不仅是点状保护的意义,更是区域保护这个系统性工程的重要环节。

　　卜内门洋行在保护中开发,在开发中保护,坚持公开性、公益性、开放性,充分发掘历史建筑潜在人文价值,恢复和保留街区古典风貌,并通过功能重整及设施更新,将改造区域打造成为独一无二的城市经典。

外滩鸟瞰图 aerial view of the Bund

In the Bund area, there are a few modern western buildings in various styles built in 1920-1936, forming the core area of the Bund Zone with Historical and Cultural Features. They are also the source of the Bund International Architecture Exhibition and modern Shanghai.

Huangpu District has basically formed a composite point-line-plane spatial system for the protection of historical features composed of "buildings with historical features - roads under protection of features - area with historical and cultural features".

Brunner Mond & Co. adjacent to the Bund building complex, is a benchmark protection and promotion project in Huangpu District. Its renovation and renewal are not of point-based protection significance, but also an important part of the systematic regional protection program.

Brunner Mond & Co. has been protected amidst development and developed amidst protection. In line with transparency, public welfare and openness, we have fully explored the potential humanistic value of historical buildings, recovered and preserve the classical style of the block, and created the renovation area into a unique urban classic through functional reorganization and facility renewal.

根据2016年发布的《上海市城市总体规划（2017—2035年）》，卜内门洋行大楼所处地块主要规划业态属科技创新型商业以及办公。修缮后的卜内门洋行大楼，功能拟设置为1层商业餐饮、5层接待展陈、其余各层高楼作为办公使用，为该区域发展注入新的活力。

作为 BA/work 系列产品中的旗舰项目，百联资控在修缮这个项目时，不是简单地复制过往经验，而是深刻地再现了原有建筑的精神，赋予了生动的商业意识形态。复新修葺后"新旧融合"，既传承了城市记忆建筑文化，又肩负了新的使命，成为兼具都会生活、精英社群办公功能，同时享有外滩极致景观的外滩城市符号。

According to the *Shanghai Master Plan 2017-2035* issued in 2016, the block in which the Brunner Mond & Co. is located is mainly planned for science and technology innovation-oriented new-type commerce and offices. The renovated Brunner Mond & Co. is planned for commercial catering on the ground floor, reception exhibition on the fifth floor, and offices on the other floors to inject new vitality into the development of the block.

The renovation of Brunner Mond & Co. is a flagship project of BA/work series products. During this project, Bailian Asset Holding has deeply reproduced the spirit of the original building and endowed with a vivid commercial ideology, instead of simply copying the past experience. The renovated building with a mix of old and new elements has not inherited the culture of an urban memorial building, but also shouldered the new mission as an urban symbol with the function of urban life and elite community offices while enjoying the ultimate landscape of the Bund.

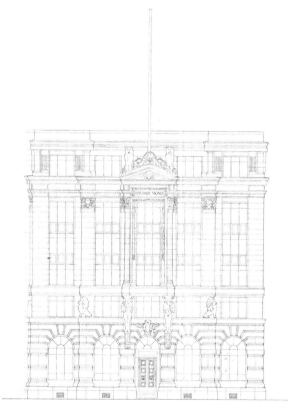

1947年东立面历史图纸 historical photo of east facade in 1947

BA/work
Project Lineup Brand Strategy

As one of the six project lineups of Bailian Asset Holding, BA/work is dedicated to making use of the optimal combinations of business forms, operating costs and personalized service concept to endow the old property assets with new vitality and generate greater value.

BA/work series office space boasts superior locations, convenient transportation, office facilities up to international first-class standards, and efficient humanistic supporting services. After fully considering the actual office needs of each location and respecting the historical development context of space, through elaborate planning and design, BA/work strives to build a communication platform for the industrial community, a win-win situation of multiple resources, and an inspirational living space and experience for the urban community.

卜内门洋行改造后外立面效果图
design sketch of the facade of Brunner Mond & Co. after renovation

BA/work 产品线品牌战略

　　BA/work 是百联资产控股公司六大产品线之一,专注于利用最优的业态组合,以最优化的运营成本、个性化的服务理念,赋予老旧物业资产新的活力,发挥其更大的价值。

　　BA/work 系列办公空间,项目位置优越,周围交通便利,拥有符合国际一流标准的办公设施和高效人性化的服务配套。**BA/work** 充分考虑各个区位的实际办公需求,尊重空间历史发展脉络,精心规划与设计,致力于打造产业社区的交流平台,缔造多方资源共赢,为城市社群打造充满灵感的生活空间和体验。

卜内门洋行现状实景图 photo of current Brunner Mond & Co.

岁月刻画，前世今生

【旧日洋行的百年变迁】

四川中路卜内门洋行是英商卜内门公司在华东地区的总部大楼，是近代上海外侨生产生活的重要例证。1949年后作为上海商业储运公司办公楼及新华书店总店使用，承载了一定的文化记忆。

四川中路南起爱多亚路（今延安东路），北至苏州路，长1270米，1855—1860年公共租界工部局辟筑，是租界中区的一条南北向干道。起初名为桥街，1865年更名为四川路。1945年收回租界后改名为四川中路，沿路多洋行机构。

位于四川中路的133号大楼始建于1921年7月，原设计人为格拉汉姆·布朗（Graham Brown）和温格罗夫（Wingrove）。二人曾参与设计怡和洋行、汇丰银行及嘉道理爵士公馆。

From the Past to the Present

Centurial Changes in Brunner Mond & Co. Building

Brunner Mond & Co. on Middle Sichuan Road is the headquarters building of Brunner Mond (UK) Limited in Eastern China and an important example of the production and life of the foreign settlers in modern Shanghai. After 1949, it was used as the office building of Shanghai Commercial Storage and Transportation Co., Ltd. and the head office of Xinhua Bookstore, bearing certain cultural memories.

The Middle Sichuan Road is 1,270 meters long, stretching from Avenue Edward VII (now Yan'an East Road) in the south and Suzhou Road in the north. It was built by Shanghai Municipal Council in 1855-1860 as a south-north main road in Secteur Central. It was named Qiao Street at first and renamed Sichuan Road in 1865. After the concession was recovered in 1945, it was renamed again as Middle Sichuan Road, along which there were many foreign institutions.

No. 133 Building on Middle Sichuan Road was built in July 1921. The original designers were Graham Brown and Wingrove, who had been involved in the design of Jardine Matheson, HSBC and Kadoorie Villa.

卜内门洋行历史照片
historical photo of Brunner Mond & Co.

唤醒 RENAISSANCE

走近这座建筑，推开铁铸大门，仿佛开启了一扇"时光之门"，百年历史积淀扑面而来。
Pushing the iron gate is like opening a "door of time". The century-old historical atmosphere overwhelms us.

* 始建于1921年7月，1923年3月14日落成，卜内门洋碱公司（世界近代史上著名的化学工业公司）将总部由今九江路迁至此，时年落成礼中西来宾到者共有300余人，英国驻沪领事也到场祝贺。

Built from July 1921 and completed on March 14, 1923, the headquarters of Brunner Mond & Co. (a famous chemical industry company in the modern history of the world) was moved here from today's Jiujiang Road. Over 300 Chinese and Western guests attended the completion ceremony, including the British Consul in Shanghai.

* 1949年，新华书店总店和上海发行所搬迁于此。21世纪初，上海新华书店总店（事业单位）仍在此办公外，组建了上海新华发行集团（企业），同在一幢楼办公。

In 1949, the Head Office & Shanghai Distribution Agency of Xinhua Bookstore was moved here. At the beginning of the 21st century, Shanghai Xinhua Bookstore (a public institution) was still here, and incorporated Shanghai Xinhua Distribution Group (an enterprise) here.

* 1956-2012年，四川中路133号大楼作为上海商业储运公司办公使用，改名为储运大楼。1975年，上海市商业储运公司设计部改造七层加层。

From 1956 to 2012, No. 133 Building on Middle Sichuan Road was used as the office of Shanghai Commercial Storage and Transportation Co., Ltd. and renamed as the Storage and Transportation Building. In 1975, the Design Department of Shanghai Commercial Storage and Transportation Co., Ltd. built another two storeys.

* 1994年，入选第二批上海市优秀历史建筑。

In 1994, it was listed in the second batch of outstanding historical buildings in Shanghai.

卜内门洋行现状实景图
photo of current Brunner Mond & Co.

卜内门，即英文Brunner Mond的汉译，取自于英国工业企业家约翰·汤姆森·卜内和发明家路德维格·门两个人的名字。英国卜内门公司（Brunner Mond & Co.）由二人合作创建于1873年，是闻名于世的纯碱与肥料的制造厂商，是ICI集团（Imperial Chemical Industries，英国帝国化学工业集团的简称）的前身。

早在19世纪中叶，比利时工业化学家索尔韦实现了氨碱法的工业化，产品由于质量纯净而被称为纯碱，民间称为"洋碱"，这是制造肥皂的主要原料。中国传统洗衣物用的是天然之物，如皂荚、稻草灰等，反映出古代农业经济特点。到了近代，西方的科学技术随着门户"开放"，源源流入中国，推动中国从农业经济过渡到工业时代。这其中，纯碱业在上海的兴起，就是在洗涤上的一次"革命"。

卜内门公司由经营洋碱起家，总部在伦敦，是英国当时的四大公司之一，在英国有300多家工厂，亦是当时世界五大公司之一。1900年，卜内门公司进驻中国，聘请在华英商李德立为总经理。

卜内门公司历史照片
historical photo of Brunner Mond & Co.

Brunner Mond came from the names of British industrial entrepreneur John Thomson Brunner and inventor Ludwig Mond, who in 1873 founded the Brunner Mond (UK) Limited, a world-famous soda ash and fertilizer manufacturer, the predecessor of Imperial Chemical Industries (ICI) Group.

As early as the mid-19th century, Belgian industrial chemist Solway industrialized the ammonia-soda process to produce a substance named soda crystals after its pure quality, also called soda ash by folks, the main raw material for making soap. Traditionally, in China, natural substances such as gleditsia sinensis and straw ashes were used for laundry, reflecting the characteristics of ancient agricultural economy. In modern times, with the opening-up, Western science and technology has flowed into China, promoting China's transition from agricultural economy to industrial era, among which the rise of soda ash industry in Shanghai was a "revolution" in washing.

Headquartered in London, Brunner Mond & Co. started its business in soda ash production. It was one of the four largest companies in the UK and one of the five largest companies in the world at that time, with more than 300 factories in the UK. In 1900, Brunner Mond & Co. entered China and hired the British businessman Edward Selby Little in China as its General Manager.

1921年7月，卜内门公司因业务兴旺，旧有房屋不够用，便在今四川路福州路口的南边买下了一块地皮，动工兴建大楼。1923年3月，7层钢筋混凝土结构的大厦建设落成，占地面积676平方米，建筑面积4636平方米，初始设计功能为办公楼，顶层局部有厨房及实验室。

卜内门上海分公司成立后，在纯碱业可谓独占鳌头，生意兴隆，因此还兼营其他业务，如化肥等产品。从20世纪初直到30年代的二十几年里业绩不断上升，名气越来越响。一直到40年代初，太平洋战争爆发，公司被日军接管。战后虽然复业，但已没有了30年代的辉煌。

1956—2012年，商业储运公司与新华书店共同使用这栋建筑期间，一层至四层为新华书店使用。一层设门卫、秘书科、食堂；二至四层分别为业务部门、发行部门和政工部门办公使用；其余楼层为商业储运公司使用，为办公及配套食堂。

In July 1921, for lack of enough premises to support its booming business, Brunner Mond & Co. bought a piece of land in the south of the interaction of Sichuan Road and Fuzhou Road. In March 1923, the seven-storey reinforced concrete structure building was completed, covering a floor area of 676 square meters and a building area of 4,636 square meters. It was originally designed as an office building, with kitchen and laboratory on the top floor.

Brunner Mond & Co. led the soda ash industry after it was established in Shanghai, so it also engaged in such businesses as chemical fertilizers. In over two decades from the beginning of the 20th century to the 1930s, Brunner Mond & Co. performed better and earned greater reputation, until it was taken over by the Japanese army in the early 1940s, when the Pacific War broke out. Although resumed after the war, the company lost its glory in the 1930s.

From 1956 to 2012, Shanghai Commercial Storage and Transportation Co., Ltd. and Xinhua Bookstore shared the building. The first to fourth floors were used by Xinhua Bookstore. The ground floor was equipped with the Guard Room, the Secretariat Division and the Canteen, while the second to fourth floors were offices of the Business Division, the Distribution Division and the Political Work Division. The rest floors were used by Shanghai Commercial Storage and Transportation Co., Ltd. as offices and the canteen.

历史资料中的四川中路133号照片
historical photo of No. 133 Middle Sichuan Road

唤醒 RENAISSANCE

【阅尽千帆的"历史中人"】

"在变幻的生命里,岁月原是最大的小偷。"这个小偷可能会带走许多珍贵的东西,然而也会留下印记,刻画进砖瓦之中,流淌于记忆之内。

距离卜内门一条街,是一片虽然陈旧拥挤,却透露着市井生活烟火气的古老城区,而这片城区也正在被现代化的新上海毫不留情地湮灭。

魏老伯在这里生活了大半生,每天天刚亮就沿着四川中路(该道路目前为滨江非机动车道,仅允许行人、非机动车及持有沿线工作证的机动车通行)遛弯,看着阳光从香樟树的叶片缝隙中投射下来,在地面上形成斑驳的光点,手里的拐杖轻轻敲击着马路,细腻而清脆的响声,仿佛历史穿透而来的回音,透着真实和纯粹。

这处承载着近代历史维持昔日孤高的城区,对于年近九旬的他来说,不是故事,不是传奇,而是生活本身。

四川中路133号夹层高窗现状
current high window of mezzanine at No. 133 Middle Sichuan Road

A Witness Living near the Brunner Mond & Co. Building

"In the changeable life, time is the biggest thief." The thief may not take away many treasures, but also leave a mark engraved in bricks and tiles, flowing into the memory.

Only a street away from Brunner Mond & Co. there is an old, crowded but mundane urban area, which is being annihilated by the modern Shanghai mercilessly.

An old man surnamed Wei has lived here for most of his life. He walks along the Middle Sichuan Road (now a riverside non-motorway, open only to pedestrians, non-motor vehicles, and motor vehicles with work permits) at dawn every day. Watching the sunlight projected from the gaps between the leaves of the camphor tree to form mottled light spots on the ground, he gently hits the road with his crutch, making delicate and clear sounds, like the echo from history, real and pure.

For him, who is nearly 90 years old, this lonely urban area which carries the modern history is neither a story nor a legend, but the life itself.

洋行内部办公场景的历史照片
historical photos of the offices in Brunner Mond & Co. Building

四川中路133号，从他有记忆开始就往来着很多外国人，他们看起来像贵族，生活十分优越，西装笔挺地进出这栋大楼，这里普通人是不能随意出入的。

No. 133 Middle Sichuan Road has been full of many foreigners since he had memories. They looked like aristocrats and led a very good life. They entered and exit the building in suits. Ordinary people were not allowed to access at will.

透过老居民楼的窗看大楼
viewing the building from the window of an old residential building

洋皂、洋碱、洋火、洋钉全部是漂洋过海进口而来，起初这些东西只有富贵人家才用得起，而这栋楼就是关于洋碱的。他隐约记得，这栋楼在抗战时期曾被日本人占领。如今这栋熬过了风云变幻的洋行大楼，倒颇有几分大隐于市的味道。

他说："你现在看好像没有了当年的模样，也觉得没有什么，可是细看啊，那些精致的华丽都隐藏在底下。立柱底层包裹的黑色花岗岩，墙角精致的浮雕和山花都在那……它是有生命的，如果能回复到它年轻的时候就好咯……"

随着魏老伯的手指，仿佛穿透百年的风雨和迷雾，时年繁盛重现于世。

正在改造中的四川中路133号正门
main entrance of No. 133 Middle Sichuan Road under renovation

Soap, soda ash, matches, and nails were all imported abroad. At first, they were only available to the rich. This building is about soda ash. He vaguely remembered that it had been occupied by the Japanese army during the Counter-Japanese War. The building has survived the war, standing like a real hermit in downtown Shanghai.
As he said: "The building seems like not the one it used to be. But if you look carefully, those exquisite elements are hidden inside. The black granite wrapped in the bottom of the columns, the exquisite reliefs and pediment… are still here. The building is zoetic, if only it could be restored to its heyday…"
His finger points as if penetrating the wind and fog of a hundred years to reappear the glory at that time.

【世纪轮回中的砖瓦诉"情"】

卜内门洋行大楼是20世纪20年代初新古典主义风格建筑在上海的重要实例。建筑细部装饰语汇丰富、手法成熟，建筑立面和室内空间效果生动。

卜内门洋行大楼是一座地上7层、地下1层的钢筋混凝土框架结构的建筑。原始功能上，地上部分除七层为设备房间、厨房及通风实验室，五层东南角有董事会会议室(board room)外，其余各层均为办公；地下室为设备用房及储藏室。

建筑外观是带有横三段式特征的新古典主义风格，显示出有节制的巴洛克建筑的影响，是20世纪20年代初新古典主义在上海的建筑实例。

建筑立面首层为具有厚重感的水平向仿石的水刷石划格，横向分隔线脚丰富而显著，东立面首层设拱形窗洞4个，拱形居中门洞1个。中部二至五层设方窗，其中三至五层中间跨设通高圆柱，柱上承三角形山花。上部六层亦设方窗，南北端两跨窗边设壁柱。

整体外墙采用青水泥进行水刷制成仿石墙面，底层外墙采用大块石砌方柱。巨大的方形倚柱置于三层与五层之间，甚为气派。下层门窗为圆拱形，具有文艺复兴的人文精神。

大楼建筑细部装饰语汇丰富、手法成熟，大门古铜色（铁质），两侧有较大的玻璃窗，窗框也用这种材质。在东立面（正立面）及南立面，墙上均设雕塑作装饰。上部挑檐及中间阳台处设圆柱，柱下也设雕塑，走廊也有浮雕，所以具有巴洛克风格特征，富丽而高雅。

此外，在原设计中，外立面二层窗间及五层山花两侧还设有男性阿特拉斯力士雕像(Atlantes，为男像装饰柱)、狮鹫、盾形纹章、翼狮等雕塑，后被人为破坏，现已缺失。

这座建筑注重细部手法，主要表现在东、南两个主要立面的交接处，从下而上过渡都很自然。无论檐部、墙身、基座，以及所有线脚，没有一处勉强，这正体现出新古典主义建筑手法的要旨理性和逻辑性，也说明设计者（英国建筑师格雷姆·布朗）的专业造诣深厚。

一砖一瓦，甚至是转角之间的衔接，都在书写那个年代独有的印记，诉说着时代的豪情。

Lofty Sentiments Hidden in Bricks and Tiles

As is an important example of neoclassical building in Shanghai in the early 1920s, the Brunner Mond & Co. boasts rich and sophisticated decorations as well as vivid facades and interior space.

The building is a reinforced concrete frame structure with seven storeys above ground and one storey underground. Originally, except for the equipment room, kitchen and ventilation laboratory on the seventh floor and the board room in the southeast corner of the fifth floor, other floors above ground were for offices; the basement was equipped with equipment and storage rooms.

With its appearance in a neoclassical style with three horizontal sections, showing the impact of moderate Baroque architectural style. The building was an example of neoclassical style in Shanghai in the early 1920s.

The ground floor of the facade features the crossing of washed granolithic plaster using heavy horizontal faux stone with abundant and significant horizontal rules. There are four arched window openings and one arched door opening in the middle on the ground floor of the east facade; square windows on the front facade from the second to fifth floors; towering columns with triangular pediments on the front facade from the third to fifth floors. The upper six floors are also provided with square windows, and pilasters between two windows in the north and south.

风姿绰约的卜内门洋行历史照片
historical photo of the Brunner Mond & Co.

The exterior wall is made of faux stone using Portland cement after washing, and the bottom of the exterior wall is made of large stone masonry columns. The huge square detached column is placed between the third and fifth floors, quite impressive. Lower doors and windows are arched, showing some humanistic spirit of the Renaissance.

The interior features rich and sophisticated decorations. The gate is in bronze (iron), with large glass windows on both sides. The window frames are also made of this material. In the east facade (front facade) and south facade, the walls are decorated with sculptures. Columns are set in the upper cornice and the middle balcony. Sculptures are set below the columns. Reliefs are also in the corridor. They embody the characteristics of Baroque style, rich and elegant.

In addition, in the original design, there were sculptures such as Atlantes (male figure used as a column), Griffin, coat of arms, and winged lion between the windows on the second floor and on both sides of the pediment on the fifth floor, which have been damaged and are now missing.

This building pays attention to the design of details, mainly demonstrated in the junction between the two main facades of the east and the south, with rather natural transition from the bottom to the top. No matter the eaves, the walls, the bases, or the architrave, there is no force, embodying rationality and logicality, the essence of neoclassical building, and showcasing the professional attainments of the designer (British architect Graeme Brown) as well.

Bricks and tiles, even the junction between two corners, are writing the unique imprint of that era, and expressing the lofty sentiments of that age.

唤醒
RENAISSANCE

外立面雕花细节现状图
photo of current carving pattern details on appearance

【精致典雅的历史笔触】

现场勘探：大楼局部房间的装饰基本保存完整，具有较高的建筑艺术价值。采用的建筑材料和工艺，经近百年使用仍然基本完好，体现了当时高超的建筑材料制作工艺。

推开卜内门洋行厚重的大门，是另一番模样，老上海那种奢靡吃重的味道褪去了，不是想象中的富丽堂皇，扑面而来的是一股克制的精致与凝练的典雅。

出入口更像通往某个幽深历史隧道的古老大门，简洁的弧线，剔透干净。加建的门卫室也并不突兀，一并融合在历史的折痕里。

穿过四叶转门走进门厅，首层高约7.6米，左右两侧因加建了夹层，将原本拱形窗户的光线压了下来，但依然挡不住它骨子里的那份大气。你甚至可以想象落成典礼的那天，宾客们眼见这素净典雅的门厅，都忍不住颔首称赞的样子。向上看，天花板上弹眼落睛的磨砂吊灯，天花板四周均勒着精巧的石膏线脚，沿着白色墙壁有一圈实木护壁，脚下则是20世纪30年代白领踏过的拼花水磨石地坪。

Exquisite Decorations

On-site survey: The decorations of partial rooms of the building are basically intact, with high architectural art value. The building materials and techniques adopted are still basically intact after nearly a hundred years of use, reflecting the superb manufacturing technique of building materials at that time.

Pushing the heavy gate of Brunner Mond & Co. is another world. Not extravagant and magnificent as imagined, the building delivers an atmosphere of restrained refinement and concise elegance.

The entrance is more likely to be a gate to a deep historical tunnel, with a simple and clean arc. The guard room built later is not abrupt, but integrated into the history.

Behind the four-wing revolving door is the foyer. The ground floor is about 7.6 meters high. Interlayers added on both sides press down the light of the original arched windows, but the magnificence of the foyer is still in its bones. You can even imagine that at the completion ceremony, the guests nodded their heads and praised the pure elegant foyer. Looking up, there is a conspicuous frosted chandelier on the ceiling surrounded by exquisite plaster lines. A circle of solid wood dado is along the white wall. At the foot is the parquet terrazzo floor that white-collar workers stepped on in the 1930s.

站在门厅看向四川中路方向的现状照片

photo of standing in the foyer to look towards the current Middle Sichuan Road

开放式井道电梯居中,两边有扶梯,扶梯侧面有腰门通至底层办公室。

室内装修颇具匠心,比如人字拼花实木地板、木质门窗套、冷弯铆接铁艺栅栏、带有西班牙风格的螺旋麻花柱式栏杆、类莲叶包、覆莲、串珠为主题雕刻精美的装饰,小细节上精雕细琢,处处有惊喜。

五楼的总经理室是最叫人惊艳的,四壁白色浑水漆涂刷,地面为实木人字拼花地板,顶棚有丰富的石膏线脚,房间西端有白色卡拉拉大理石为主色调、黑框米黄色大理石装饰的假壁炉。地板、木护壁、天花、壁炉等装饰构件保存状况良好。但禁不住岁月的侵蚀,局部墙面与天花涂料已起翘、开裂、脱落。

大楼顶部精美的装饰细节
delicate details of decoration on the top of the building

In the middle, there is an open shaft elevator, with escalators on both sides. A wicket door is set beside each escalator to the office on the ground floor.

The interior decoration is ingenious, evidenced by the exquisite and amazing herringbone parquet solid wooden floor, wooden door and window covers, cold-roll riveting iron fence, spiral mahogany column handrail in Spanish style, lotus leaf-like bag, lotus petal and a string of beads.

The General Manager's Office on the fifth floor is the most stunning. There are walls painted with whitewater-borne coating, herringbone parquet solid wood floor, rich plaster lines on the ceiling, as well as a false fireplace with white Carrara marble as the main material and supplemented by yellow marble with black frames in the west corner of the room. Decorative components such as floor, wooden dado, ceiling, and fireplace are well preserved. Unfortunately, due to the erosion for years, part of the walls and ceiling coating have been warped, cracked, and fallen off.

保存完好的入口挑空大厅
well-preserved Atrium Hall at the entrance

总经理室壁炉造型现状
photo of the current false fireplace in the General Manager's Office

五层大厅现状照片
photo of the current lobby on the fifth floor

近百年的时光也在它身上留下了每段时光的印迹，后期加设的吊顶、铺地、隔墙、现代感的灯具、装饰物，不免有着些许不相称的违和感。

循古朴的木梯而上，从顶层的窗口望出去，绚丽的霓虹灯凸显着建筑群的辉煌，淋漓尽致地渲染着上海华贵、浪漫、大气的别样风情，黄浦江上耀眼的光芒有一种近代的沧桑。

简言之，现存建筑有一副好"骨架"，但急需赋予新的血肉和能量。

时至今日，复新的不仅仅是一所洋行，更是赋予了多功能的科技创新型商业及办公大楼的憧憬。于是，怎样在保留建筑历史遗迹，维护建筑原始价值前提下加入现代元素，满足现时的功能需求和未来的可持续发展理念，成为建筑师首要思考的问题。

第一章　CHAPTER ONE

Nearly a hundred years have passed, leaving a mark of every period of time on it. The ceiling, flooring, partition as well as modern lamps and ornaments added later are inevitably somewhat incongruent.
Upstairs the wooden ladder, looking out by the window on the top floor, the gorgeous neon lights highlight the brilliance of the building complex, vividly rendering the unique style of Shanghai, luxurious, romantic and magnificent. The dazzling rays on the Huangpu River have a sense of modern vicissitudes.
In short, the existing building has a superb framework, but it is in urgent need of new blood and energy.
Today's renovation of this building is more to turn it into a multi-functional science and technology innovation commercial and office building. Therefore, how to add modern elements to meet the current demand for functions and the future sustainable development concept while preserving the historical relics and maintaining the original value of the building has become the first consideration of architects.

窗外的上海城市风景
city scenery outside the window

唤醒
RENAISSANCE

大楼顶部现状俯视图
the top of the building

Superb Framework and New Soul

Inheritance of the History and Integration into the Modern Time

Brunner Mond & Co. is a true cross-section of history and a miniature of China in the old era. Its replacement of functions is an important exploration case of Shanghai's urban renewal and development, which is of positive social significance and promotion value.

According to the *Notice on Protection Requirements for Renovation Project of the Outstanding Historical Building at No. 133 Middle Sichuan Road* (HLB [2018] No. 45) issued by Shanghai Historical Building Protection Affairs Center: "The protection requirements for No. 133 Middle Sichuan Road are of Class II and according to Article 25 of the *Regulations of Shanghai Municipality on the Protection of Places of Historical and Cultural Importance and Outstanding Historical Buildings*, the facade, structural system, basic plane layout and characteristic interior decorations of the building shall not be changed."

好"骨架"与新"灵魂"

【传承历史与共融共生】

卜内门洋行是历史真实的横截面,是旧时代中国的一个缩影。本次功能置换后,是上海城市更新和发展的重要探索案例,具有积极的社会意义和推广价值。

上海市历史建筑保护事务中心《关于四川中路133号优秀历史建筑修缮工程保护要求告知单》(沪历保【2018】45号)中要求:"四川中路133号的保护要求为二类,根据《保护条例》第二十五条,建筑的立面、结构体系、基本平面布局和特色的内部装饰不得改变。"

根据保护管理技术规定,四川中路133号的东立面为外部重点保护部位,进厅、门厅、大厅的空间格局、五层总经理室及其他原有特色装饰等为内部重点保护部位。修缮前应认真考证原始设计资料及施工工艺等,重点保护部位应严格按原式样、原材质、原工艺进行修缮。

内部历史照片
historical photo of the interior

According to the *Technical Provisions of Shanghai Municipality on the Urban Planning Administration*, the east facade of Brunner Mond & Co. is a key exterior protected part, while the entrance hall, the foyer, the spatial pattern of the lobby, the General Manager's Office on the fifth floor and other original characteristic decorations are the key interior protected parts. Before repair, the original design data and construction technology shall be carefully verified, and the key protected parts shall be repaired in strict accordance with the original style, materials and techniques.

由于该项目历经百年岁月，几番易主多次改造，现状与落成初始已有较大区别。设计团队多方调阅历史档案，深入剖析历次改建改造情况，结合现状、原始设计图纸照片和本次改造目标，审慎地制订了项目的保护修缮方案，形成了三条修缮设计目标，这也成为项目后续改造的根基和指引：

* 目标一 提高完整性：大楼现状保存较好，但由于历年功能调整，外立面附着物较多，各层室内重点保护空间风格不一。本次修缮依据优秀历史建筑的修缮原则，从大楼的建筑风格、内部格局、使用功能和装饰细部等方面出发，综合考证，改善提高历史建筑的整体完整性，提升历史建筑的文化价值。

* 目标二 提高安全性：从分析大楼的历史资料及现有状况着手，审慎策划新功能，通过现代计算手法、设计应用、施工工艺等各渠道努力，提升历史保护建筑的结构与构件安全性。

* 目标三 提升环境舒适度：根据使用功能需求，遵照在保护的前提下，尽量满足节能高效的设计原则，选择先进适用的设施设备，整体提升历史建筑的内在使用品质。

2019年4月，上海市历史建筑保护事务中心领导和来自规划、建筑和施工方面的专家评审组对项目进行了现场踏勘，考察建筑保护现状，认真勘察每一层格局、装饰风格、外立面细节部位及雕塑情况。随后，专家们召开了保护修缮方案预征询专家评审会，原则同意设计单位的方案，建议进一步深化历史考证，恢复原有历史信息及室内格局，原样恢复外立面细部、门厅、电梯间、室内装饰等的设计风格，关于外立面雕塑还应符合历史原貌。

四川中路133号经过这次评审，在接下来的修缮中将改变相应的使用功能，提高建筑结构强度。在施工中注重它的历史文化和周边环境协调，对它进行恢复性修复，使之达到缺失部分的修补与整体保持和谐。在保留其特性基础上增添一些新的元素进行创新性开发，建设成为城市新景观，让古建筑与新建筑融合为一体，创造出新的价值。

外立面实景图
photo of appearance

1947年版历史图纸
Historical Drawing in 1947

历保专家对
修缮方案的评审意见
review opinion of protection experts of historical buildings on the repair plan

This century-old building has been renovated several times by its owners. The status quo is quite different from the one initially completed. Upon review of historical archives and in-depth analysis of previous reconstruction and renovation situations, coupled with status quo photos, original design drawings and the renovation objective, the design team carefully formulated a protective repair plan with three design objectives as the foundation and guidance for the subsequent renovation of this project:

* Objective 1: Improve the integrity: The building is well preserved, but due to functional adjustments over the years, there are many attachments on the facade, and the interior key protected spaces on each floor are in diverse styles. In line with the repair principle of outstanding historical buildings, through a comprehensive investigation, this repair focuses on the building style, interior pattern, use functions and decoration details to improve the overall integrity and cultural value of this historical building.
* Objective 2: Enhance the safety: Upon analysis of the historical data and status quo of the building, new functions are carefully planned to enhance the safety of the structure and components via modern calculation methods, design applications, construction technology, etc.
* Objective 3: Raise the environmental comfort: According to the demand for use functions, on the premise of protection, the design principles of energy saving and efficiency is prioritized with advanced and applicable facilities and equipment to raise the internal use quality.

In April 2019, the leaders of Shanghai Historical Building Protection Affairs Center and the review team composed of planning, architecture and construction experts conducted an on-site survey on the protection status quo of the building, including the pattern of each floor, decoration style, details of the facade and sculptures. Then the experts held a pre-consultation review meeting and agreed on the protective repair plan put forward by the design entity in principle. They suggested further deepening the historical research, restoring the original historical information and interior pattern, resuming the design styles of the facade details, foyer, elevator room, interior decorations, and imitating the historical original appearances of the facade sculptures.
After the review, the corresponding use functions of Brunner Mond & Co. would be changed to improve the structural strength. In the construction, the coordination between its historical culture and the surrounding environment is valued for restorative repair to make the renovated missing part harmonious with the whole. New elements are added while retaining its characteristics for innovative development to integrate the new and old buildings into a new landscape of the city, and create new values.

楼梯铁艺护栏现状图
photo of current iron guardrail of stairs

【新与旧的活力交汇】

修缮方案表明了老建筑的历史DNA将被妥善保留修复，作为一种文化遗产植入改造一新的功能空间中，让现代人也可以品读这份传承了百年的典雅情韵。

2019年盛夏，整栋大楼迎来了百年来最彻底的一次整修与翻新。结合实际，立足长远，着眼于传承历史，保护性开发的原则，设计团队围绕"拆、改、留、修"对这栋在历史中浮沉、充满故事的老建筑开出了一张"处方"。

外立面及室外修缮内容包括：

* 拆除外立面空调支架、无用管线等附着物。
* 清洗修缮外立面水刷石饰面，修复缺失与空鼓部分。
* 保留修缮现有实腹钢窗及入口木门，修缮缺失窗扇、五金件等部件，结合钢窗型材尺寸和窗扇承载能力，考虑实腹钢窗玻璃更换为6-9a-6中空玻璃的技术可行性。木质门窗脱漆粉刷。按历史式样更换残损外门窗。
* 按南立面现存实样，翻制东立面缺失的黑色金属浮雕。
* 根据历史图纸及照片，复原入口处上方盾形花饰。
* 更换破损的落水管。
* 整饬西侧背巷，拆除品质较差的违建棚屋。

室内装饰修缮内容包括：

* 修缮进厅、门厅、首层楼电梯厅，利用现有地下室空间作为电梯基坑，更换电梯，以1小时耐火耐热玻璃封闭楼内电梯井镂空铁艺部分。
* 修缮五楼总经理室。
* 修缮各楼层的电梯厅格局。
* 拆除局部夹层楼板，恢复沿街历史风貌。
* 保留及修缮其他部位的木质门窗套、实木拼花地板、水磨石地坪、实木护壁等特色装饰。

百年老建筑的元素和特色部位被妥善地保留和修缮，历史文化氛围得以传承，卜内门洋行形成独特的空间体验，将会给未来的用户带来耳目一新的感觉。

Dynamic Convergence between New and Old Elements

The renovation plan shows that the historical DNA of the old building will be properly preserved and repaired. As a cultural heritage, it will be implanted into a new functional space, so that modern people can appreciate the elegant charm that has been passed on for a hundred years as well.

In the midsummer of 2019, the building ushered in the most thorough renovation in a century. Based on the reality, taking a long-term view, and focusing on the principle of historical inheritance and protective development, the design team renovated this old building full of stories in the history from demolition, modification, retention and repair.

Renovation of the facade and the exterior:

*Demolish the attachments such as air conditioning bracket and useless pipelines on the facade.
*Clean and repair the washed granolithic plaster overcoating, and the missing and hollowing parts on the facade.
*Retain and repair the existing solid web steel windows and wooden entrance door, repair the missing window sashes, hardware and other components, and consider the technical feasibility of replacing the solid web steel window glass with 6-9a-6 hollow glass in combination with the size of steel window profile and the bearing capacity of the window sashes. Repaint the wooden door windows. Replace the damaged exterior door windows in their original styles.
*Duplicate the black metal relief missing on the east facade according to that on the south facade.
*Recover the coat of arms above the entrance according to the historical drawing and photos.
*Replace the damaged downpipe.
*Rectify the back lane on the west, and dismantle the low-quality illegal sheds.

Renovation of the interior decorations:

*Repair the foyer, the lobby, and the elevator hall on the ground floor; replace the elevators with the existing basement space as the foundation pit and seal the hollow-out iron part of the elevator shaft with glass resistant to fire and heat within one hour.
*Repair the General Manager's Office on the fifth floor.
*Repair the pattern of the elevator hall on each floor.
*Dismantle partial mezzanine and restore the historical features along the street.
*Retain and repair the featured decorations such as wooden door window sets, solid wood parquet floor, terrazzo floor, and solid wood dado on partial parts.

With elements and featured parts properly preserved and repaired to inherit the historical and cultural atmosphere, Brunner Mond & Co. will bring a refreshing unique space experience to future users.

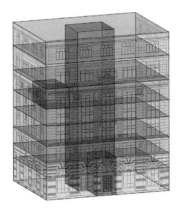

室内修缮重点保护部位
key interior protected parts

经典邂逅与价值再生

【时代的映射：墙面雕塑】

立面墙面

外墙面以水刷石作为墙面饰面，底层的外墙面仿大块石砌式样，五扇拱形门窗间以六尊敦实厚重的柱基础形式落地。外墙现状斑驳，有严重的风化受潮现象，外空调机严重影响立面形象。

基于外立面保存情况较为完好的现状考察，对于外立面的修复主要集中在拆除无用的附着物、清洗修缮水刷石饰面，修复缺失与空鼓部分。对窗扇、木门、五金件部分也按历史原样式进行修复、脱漆粉刷或更换残损部分。

外部立面及门窗现状
current facade and exterior door windows

Encounter with Classics and Rebirth of Value

Mirror of the Time: Wall Sculptures

Facade Wall

The exterior wall is decorated with washed granolithic plaster, while the exterior wall of the bottom is in quasi-big-stone style. Five arched door windows fall to the ground through six heavy column foundations. The exterior wall is mottled due to serious weathering and damp, and the outer air conditioning has a serious impact on the facade image.

Based on the status quo investigation, the facade is kept intact, so restoration focuses on removing useless attachments, cleaning and repairing the washed granolithic plaster overcoating, repairing the missing and hollowing parts, as well as renovating, repainting or replacing the window sashes, wooden doors and hardware parts according to their original patterns.

立面罗马柱现状图
photo of current marble pillars on facade

立面雕塑

改造的难点在于外立面雕塑部分的修复。

在卜内门大楼原来的建筑上,第六层和第七层的中间部分是凹的,在三楼和五楼的柱子之间还有各种各样的人物浮雕,柱的上方是一个希腊三角形屋顶浮雕,在第六层檐下是传统的狮子雕像滚动绣球。其中翼狮、阿特拉斯男像柱、狮鹫、盾形纹章等雕塑,据载于1965年拆除。

黑色金属浮雕莨苕花环在南立面仍存在四个较完好的实样,得以作为复制东立面上的缺失莨苕花环的依据。

南立面黑色浮雕莨苕花环现存实样
photo of the extant ferrous metal acanthus wreath relief on the south facade

东立面黑色浮雕莨苕花环历史照片
photo of the historical ferrous metal acanthus wreath relief on the east facade

Facade Sculptures

The renovation difficulty is the repair of sculptures on the facade.

Originally, the part between the sixth and seventh floors was concave; on the third and fifth floors, there were various figure reliefs between the columns, above which was a Greek triangular roof relief; under the eaves of the sixth floor was a traditional lion statue rolling a ball. Sculptures such as winged lion, Atlas, Griffin, and coat of arms were reportedly demolished in 1965.

There are still four intact ferrous metal acanthus wreath reliefs on the south facade, which can be used as the basis for copying the missing ones on the east facade.

【狮鹫】

东立面及南立面二层檐下原有共4只狮鹫雕塑,现已缺失。该处雕塑历史照片与历史图纸出入较大,拟以历史照片为基准进行复原。

[Griffin]

There were four Griffin sculptures under the eaves of the second floor of the east and south facades, which are missing now. Due to the large discrepancy between the historical photo and the historical drawing, the historical photo is preferred as the basis of restoration.

立面侧坐狮鹫左视与右视历史照片
historical photos of a sitting griffin sideways on the facade in the left view and right view

【翼狮】

在原卜内门洋行正立面入口上方，挑阳台两侧树立两根贯通三层、柱头复合柱式的圆柱，柱上承三角形山花及贝状装饰，原山花两侧各有翼狮一头。翼狮图案曾出现在古代苏美尔文明及波斯文明的装饰中，也是中世纪诸多家族纹章的常用图案。在威尼斯共和国的徽章中，翼狮是共和国守护神圣马可的纹章象征。现花饰、旗杆尚存，但两头翼狮早已缺失。

翼狮及山花历史照片与现状照片对比
historical photo and current photo of winged lions and the pediment

[Winged Lion]

Originally, above the entrance of the front facade, a column with a composite head running through three floors was placed on each side of the cantilevered balcony. The columns bore a triangular pediment and shell-shaped decorations. There was a winged lion on each side of the pediment. The winged lion pattern was seen in the decorations of ancient Sumerian civilization and Persian civilization, commonly used in many family heraldries in the middle ages. In the emblem of the Republic of Venice, the winged lion is the symbol of the patron saint San Marco. The two winged lions are missing, only leaving the floriation and flagpoles.

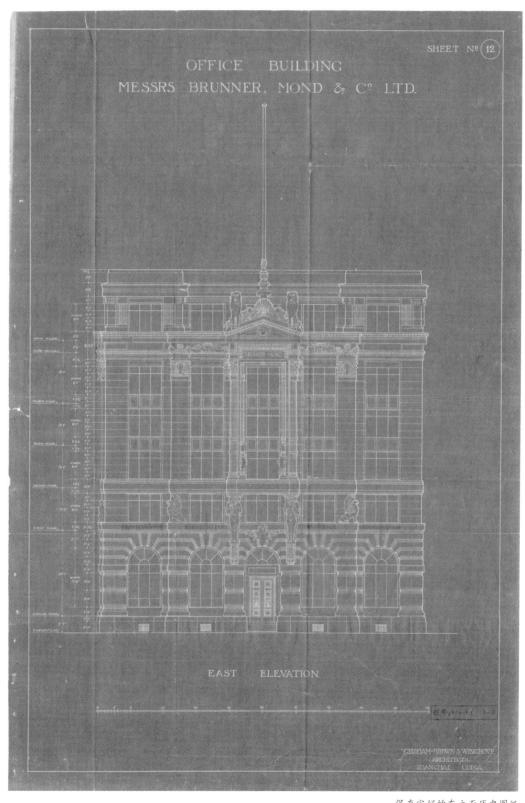

保存完好的东立面历史图纸
historical drawing of the well-preserved east facade

【扛柱力士阿特拉斯】

古典主义建筑风格流行以男性雕像来装饰建筑中的立柱、桥台和支柱，这种支撑性的结构被称作阿特拉斯。阿特拉斯（希腊语 Ἄτλας，英语Atlas）是希腊神话里的擎天神，属于泰坦神族，被宙斯降罪用双肩支撑苍天。

原建筑东立面正中三至五层有罗马混合柱式塑角柱装饰门廊，柱下原有男性阿特拉斯雕像，两侧雕像姿势不一，脚踏毛茛叶状柱托。两力士中门楣处有盾形纹章装饰，现亦缺失。

扛柱力士阿特拉斯历史照片
historical photo of Titan Atlas

[Titan Atlas]
In classical building style, columns, abutments and pillars are often decorated with male statues. This supporting structure is called Atlas. Atlas (Ἄτλας in Greek) is a Titan condemned by Zeus to hold up the celestial heavens with his shoulders in Greek mythology.
Originally, a corner column in Roman composite order stretching from the third to fifth floors was placed to decorate the portico; under the column were two male Atlas statues in different postures, each stepping onto a buttercup-shaped base. There was a coat of arms on the doorway between them, but is also missing now.

最初的设想是复原这几组雕塑，但是经过资料考证和实地勘验后发现，复原雕塑需要增加底部承重结构，但是对于历保建筑来说，外立面严格规定不可添加任何附着结构物。因此考虑到安全因素，不得不放弃对这两尊雕塑的复原。本次修缮后在使用过程中，拟通过出版物、展览、AR增强现实、节庆投影等手段，留存传播这一历史信息。虽然这是一个遗憾，但也是为了更好地保护历史建筑而做的合理取舍。

扛柱力士位置现状照片
photo of status quo Titan Atlas

The original idea was to restore these sculptures, but after data investigation and field inspection, to restore sculptures, a bottom bearing structure must be added. For protective historical buildings, however, the facade is strictly prohibited from adding any attachments. Therefore, out of safety, the restoration of the two sculptures had to be given up. In the process of use after repair, this historical information would be retained and disseminated through such means as publications, exhibitions, AR, and projection during festivals. This is a pity, but it is a reasonable choice for better protection of historical buildings.

【世纪之门：券门】

券门也可以称为"拱券门"。拱券一般都由砖石砌筑，所以券门也就是指用砖石砌成的半圆形或弧形的门洞。

卜内门洋行大楼正门外墙仿制石砌券形拱门，底楼大门两侧也为券形、大落地玻璃窗。

拱券门的形状非常优美，门洞边缘即券脸位置装饰了雕刻图案，更增添艺术性与观赏性。

拱券门现状照片
photo of the current arch gate

古铜色铸铁大门上的铜制把手是两个狮子衔环，很有分量，有一种不怒自威的庄严。进入门厅，门楣上的五角星是时代刻下的烙印。这些岁月的痕迹都很好地保留了下来，在修复过程中，重在还原，重在历史精髓的重现。

Gate of the Century: Arch Gate

An arch gate is also known as a vaulted gate. Arch is generally built of masonry. Therefore, an arch gate is a semicircle or arc-shaped gate opening built of masonry.

The exterior wall of the front gate of the Brunner Mond & Co. is an arch gate made of faux stone. On both sides of the gate on the ground floor are also equipped with large floor-to-ceiling arch glass windows.

The arch gate is well-shaped. The arch front on the edge of the gate is decorated with carved patterns, adding artistry and appreciation.

The copper-made handles on the bronze cast iron gate are two heavy and august lions each with its ring pull. The five-pointed star on the lintel of the lobby is the imprint of the times. These traces have been well preserved. In the process of restoration, the focus is on the reappearance of the essence of history.

第一章 CHAPTER ONE

饱含经典印记的建筑图腾
classic decorations on the building

【锁住老故事：窗】

20世纪二三十年代，上海堪称东方的国际大都市，华洋交处，各个国家各个地方的文化在上海交融汇集，孕育出了集优雅、浪漫、摩登于一体的老上海情调，既有东方的婉约意境，又借鉴了西方简单的生活方式。

小小的一扇钢窗，也是这种老上海情调的表达。修缮每个楼层的窗，最重要的主旨就是在保证功能性的基础上维持这一份情调。为了这份情调，修缮又增加了一定的难度。

金属基层油漆主要用于铸铁栏杆、钢窗、金属饰面板等金属基层构件的油漆保护。凡重新油漆部位，均需首先检查构件原有漆膜剥落老化情况，根据实际情况起底出白后，重新批嵌打磨。脱漆出白后，所有外露铁件均应刷防锈漆二度，调和漆三度，黑色面漆。金属件接缝要严密，用于室外的金属件接缝处用树脂涂料二道密封。

对现存尚好的钢门窗予以除锈、校正，修复后重做防锈漆及黑色面层漆，露面铁件做防锈底漆一道面漆二道，不露面铁件做二道防锈漆。对严重锈蚀无法使用的钢窗，则需要按照原门窗形式重新定制予以更换。对于玻璃破损或风化的钢门窗，利用原窗框，按原尺寸更换为中空玻璃。

五金件修缮前应委托专业单位对现有五金件进行金相分析。

修缮中应完整保留、清洗除锈、修复历史原执手、撑杆、合页、门锁等铜五金件。凡需拆卸修复的五金件应编号保护，修复后原件、原位安装，恢复历史原貌。部分缺失或损坏的门窗执手、风钩等五金件要按照相邻门窗完好五金件进行仿制。

通高拱窗现状
photo of the current full-height arch window

程大利女士在仔细考查每扇钢窗及五金件
careful check of steel window and hardware by Ms. Cheng Dali

夹层高窗现状
current high window of mezzanine

Sealing the Old Story: Windows

In the 1920s and 1930s, Shanghai was a true international metropolis in the Orient. With domestic and foreign civilizations converged here, Shanghai has bred an old Shanghai sentiment that integrates elegance, romance and modernity, combining the Oriental implication and the Western simplicity.

Even a small steel window can be an expression of such sentiment. To repair the windows on each floor, the most important thing is to maintain this sentiment on the basis of ensuring functionality, which has increased difficulty.

Metallic paint is mainly used for paint protection of metallic components such as cast-iron railings, steel windows, and metal veneers. For parts to be repainted, it is necessary to first check the peeling and aging of the original paint film of the component and bare the metal according to the actual situation before patching polish. After baring the metal depainting, all exposed iron parts shall be painted with two coats of antirust paint, three coats of ready mixed paint, and finally black finishing coat. The joints of metal parts shall be tight, and the joints used for outdoor metal parts shall be sealed with two coats of resin emulsion paint.

The well-preserved steel door windows shall be derusted and corrected. After repair, the antirust paint and black finishing paint shall be redone. The exposed iron parts shall be provided with one coat of antirust primer paint and two coats of finishing paint, while the non-exposed iron parts shall be provided with two coats of antirust paint. Steel windows severely rusted and cannot be used shall be replaced according to the original patterns. Steel door windows with damaged or weathered glass shall be replaced with hollow glass using the original window frames.

Before repairing the hardware, a professional institution shall be entrusted to carry out a metallographic analysis.

Copper hardware such as handles, brace rods, hinges, and door locks shall be completely reserved, cleaned, derusted, and repaired. Hardware to be removed for repair shall be numbered for protection and after repair, the original parts shall be installed in situ to restore their original appearances. Some missing or damaged hardware such as door window handles and air hooks shall be modeled on the intact hardware of adjacent door windows.

【时间的背面:首层进厅及门厅】

门厅空间格局现状较历史原状改动较小,除进厅加建门卫室,电梯厅因后期电梯改造垫高三级台阶之外,整体空间格局仍维持原貌。

为了恢复进厅空间的完整性,加建的门卫室将被拆除,而墙面、老门扇、天花装饰线脚等历史原物都将会得到专业人员的修缮。灯具也将被更换成更能提升门厅空间氛围、又与老建筑格调一致的式样。

进厅现状照片
photo of the current foyer

Reverse Side of Time: Foyer and Lobby on the Ground Floor

The spatial pattern of the lobby has few modifications compared with the original one, only with a guard room added to the foyer and three steps raised to the elevator hall due to the later elevator renovation.
In order to restore the integrity of the foyer space, the guard room built later would be removed, and the original objects such as walls, old door leaves, and ceiling decoration lines would be repaired by professionals. Lamps would also be replaced in a pattern that can enhance the atmosphere of the lobby space and is consistent with the style of this old building.

入口门厅现状图
photo of current lobby at entrance

原设计中，电梯基坑深度约0.9米，更新电梯时，由于原基坑深度不足，将电梯厅区域垫高了三级台阶高度。根据历史图纸及现场初步勘测，可利用部分地下室空间作为电梯基坑使用，使得基坑深度最大处达到2.6米，从而既可满足目前电梯安装的需求，又可使电梯厅地面与门厅地面平接，让门厅空间得以延伸，更加敞阔通透。

根据历保要求，首层电梯厅墙面将更换为木饰面设计，取代现有的黑色石材墙面，保持全楼风格的完整性。除夹层木饰面局部保留原有线脚之外，其余部分木饰面材质古典，细节现代，营造出典雅的时尚感。

电梯门采用古铜材质，使电梯墙低调融入门厅氛围，成为历史空间的背景。门厅增设侧喷，保证消防措施，并新增空调地柜机，提升门厅的舒适性与安全性。

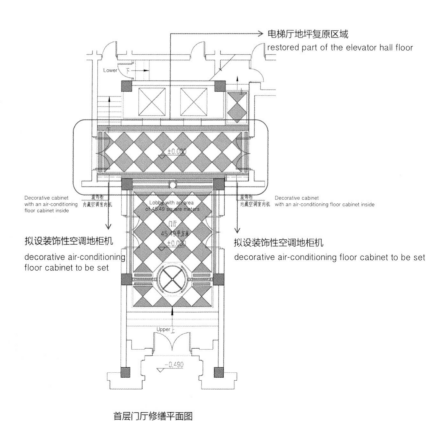

首层门厅修缮平面图
repair plan of the lobby on the first floor

首层门厅修缮效果图
repair design sketch of the lobby on the ground floor

Originally, the elevator foundation pit was about 0.9 meters deep. In the renewal of elevators, the elevator hall is padded up by three steps due to the substandard depth of the original foundation pit. According to historical drawings and the preliminary on-site survey, part of the basement space can be used to deepen the foundation pit up to 2.6 meters, which can not meet the needs of the current elevator installation, but also realize the abutment joint between the ground of the elevator hall and the ground of the lobby, so that the lobby space can be more spacious.

According to the protection requirements of historical buildings, the existing black stone veneer wall of the elevator hall on the ground floor would be replaced with a wood veneer wall to maintain the integrity of the whole building style. In addition to the original architrave retaining in some parts of the wood mezzanine veneer, the other parts use classical materials with modern details, creating a sense of elegant fashion.

The bronze elevator hall makes its wall silently mix into the style of the lobby as part of the background of historical space. In the lobby, a side spray is added to ensure fire-fighting measures and an air-conditioned floor cabinet is added to improve the comfort and safety of the lobby.

【找回昔日旧貌：一层夹层】

一层层高约7.645米，在原设计中，南北两侧使用空间应为通高空间，但历史上一层已建有夹层。根据历史图纸与历史照片推断，20世纪四五十年代，原夹层仅加至沿街面后退一柱跨位置，东立面存在一跨通高空间，现状夹层为满铺。

1921年建筑历史图纸，剖面图中首层为通高空间
historical drawing in 1921,
with floor-to-ceiling space on the ground floor

从20世纪40年代的照片中可看到一层至夹层的室内楼梯
photo of staircase between the ground floor and the mezzanine in the 1940s

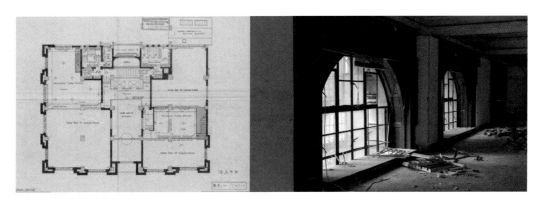

1951年历史装饰图纸夹层平面中可看到一层至夹层的楼梯位置
location of stairwell between the ground floor and the mezzanine on a historical mezzanine decoration drawing in 1951

一层夹层上部和下部的东立面通高拱窗现状照片
photo of the current floor-to-ceiling arch window between the mezzanine upper and mezzanine lower on the ground floor on east facade

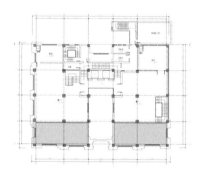

修缮设计中夹层拆除范围
removal range of the mezzanine in repair design

现有夹层楼面加建于1951年之后，局部影响沿街历史风貌，且该结构形式影响全楼耐火等级。为了实现沿街商业界面品质提升，尊重历史原貌，本次修缮中拟将夹层木格栅楼面更换为轻钢楼面，局部拆除106.4平方米夹层楼板，打造沿街通高空间，恢复重点保护部位——东立面沿街的历史原貌。

夹层局部拆除整治后，首层、夹层的室内装饰以及首层与夹层之间的楼梯，拟参考历史照片及局部现存实物进行修缮与复原设计，重现历史风貌。

Back to the Past: Mezzanine on the Ground Floor

The ground floor is about 7.645 meters high. In the original design, the space in the north and south was floor-to-ceiling space, but there was a mezzanine on the ground floor. According to historical drawings and historical photos, in the 1940s and 1950s, the original mezzanine was only extended another column along the street, and there was a floor-to-ceiling space in the east facade, and the current mezzanine was wall-to-wall.

The extant mezzanine floor was built after 1951, which partially affects the historical features along the street and the fire-resistance rating of the building. In order to improve the quality of commercial interface along the street and respect the original appearance, by replacing the wood grid floor of the mezzanine with the light steel floor and removing 106.4 square meters of mezzanine floor, a floor-to-ceiling space along the street will be created to restore the original appearance of historical features on the east facade, which are a key protected part.

Upon partial removal and renovation of the mezzanine, the interior decorations of the ground floor and the mezzanine, as well as the stairs between them will be repaired and restored according to historical photos and local extant objects in order to reproduce the historical features.

悬挂有上海商业储运有限公司门牌的照片
photo of the nameplate of Shanghai Commercial Storage and Transportation Co., Ltd.

【空间的诗学：总经理室】

位于五楼的总经理室是此次修缮的重点空间，过去这里被人称为"大班室"。"大班"一词是粤语中日常口语词，最初用于描述19世纪末20世纪初在我国的外国商人。鸦片战争前，外国商船中的管货和处理商务的货长，依照其职务高低，分别称为大班、二班、三班。以后逐渐推广开来，就成为对洋行经理的称呼，一度泛指"富豪""商贾"或"买办"等在当时有一定社会地位的人群，进而演化为市民口中的"老板"。

东侧窗外一览黄浦江景色
scenery of Huangpu River by the east window

Poetics of Space: General Manager's Office

The General Manager's Office on the fifth floor is a highlight of this renovation. In the past, it was called the Tai-Pan's Office. Tai-Pan is a Cantonese term describing foreign businessmen operating in China in the late-19th century and the early 20th century. Before the Opium War, the captains in charge of cargos and business affairs in foreign merchant ships were respectively called Tai-Pan, Secondary Tai-Pan and Tertiary Tai-Pan according to their positions. Later, it referred to a leader of a foreign firm. It was used to represent people with certain social status at that time, such as "the rich", "merchants" or "compradors", and gradually, it became the "boss".

总经理室全景现状图
panoramic photo of current General Manager's Office

总经理室现状照片 photo of the current General Manager's Office

卜内门洋行"大班室"的故事，和李德立有关。

爱德华·赛比尔·利特尔（Edward Selby Little），中文名"李德立"，生于英国多塞特郡温特波恩金斯顿，曾经就读于剑桥大学。1886年，与妻子来华传教。1900年，他辞去了教会职务，转行经商，出任上海英商卜内门公司（Brunner Mond & Co.,Ld.）东方总号总经理。

他一生的履历着实丰富，关于经营卜内门洋行的这一段，在1908年时西人评价道："卜内门东方业务由总经理李德立于1900年开始经营，业务量在他的领导下稳步增长。"

李德立画像
portrayal of Edward Selby

现在，我们看到的总经理室以白色为基调，四周实木护壁现为白色混水漆涂刷。然而在查阅历史档案和现场考证后，根据本楼其他楼层历史照片以及同时期其他楼栋相近做法推断，原貌应为木原色护壁。

同一设计师格拉汉姆·布朗参与设计的外滩三号怡和洋行会议室照片
photo of the conference room of Jardine Matheson & Co. at No. 3 on the Bund designed by the same designer Graham Brown

The story of Tai-Pan's Office in the Brunner Mond & Co. has something to do with Edward Selby Little.

Edward Selby Little, born in Winterbourne Kingston, Dorset, England, was graduated from the University of Cambridge. In 1886, he came to China for preach with his wife. In 1900, he resigned from the church and served as the General Manager of Brunner Mond & Co. in Shanghai branch.

He led a rich life. In 1908, Westerners commented on his management of Brunner Mond & Co.: "In 1900, the business of the Oriental branch was started by the General Manager Edward Selby Little, under whose leadership the business volume grew steadily."

Today's General Manager's Office is mainly in white, with solid wood dado around painted with white mixed water coating. However, upon consulting the historical archives and on-site investigation, according to the historical photos of other floors of the building and the similar practices of other buildings in the same period, its solid wood dado is inferred to be wood-colored dado at that time.

唤醒
RENAISSANCE

五层总经理室修缮效果图
renovation design sketch of the General Manager's Office on the fifth floor

本次修缮中拟保护修缮该房间的装饰，其中实木护壁进行脱漆修缮，还原为原木色，重现历史风貌。修缮一新后的总经理室，将更显出复古且优雅的魅力，在柔和灯光下泛出一层持重雅致的韵味，等待着跨越近一个世纪后再次目光的投赏。

总经理室地板及花形装饰
floor and flower decoration in the General Manager's Office

In this renovation, protective repair shall apply to the decorations in this room, in which the solid wood dado will be repainted the wood color to reproduce its historical feature. The renovated General Manager's Office will highlight ancient and elegant charm. Under the soft light, its beauty of dignity and elegance will attract visitors again after nearly a century.

【优雅的驻足：各楼层电梯厅】

历史图纸中各层楼电梯厅同中存异，其中二层、三层格局相同，四层、五层楼电梯厅装饰手法、门窗样式与其他各层相同，但电梯厅尺寸、门窗细节略有差异。

现状电梯厅除二层的格局、样式与历史图纸几乎相同之外，其余各层均有改动。各层楼梯半平台处的单扇木门据推测仍为原物。

本次修缮设计中，拟根据各层实际情况，以历史样式为参考，复原各层电梯厅空间格局，重现各层电梯厅历史风貌。

二楼电梯厅现状照片
photo of the current elevator hall on the second floor

四楼电梯厅现状照片
photo of the current elevator hall on the fourth floor

根据历史图纸及测绘图纸来看，三层、四层电梯厅空间格局略有改动，但现状风貌比较协调，除局部门窗洞口需参照历史样式进行调整外，其余室内装饰均按现状保留修缮。

五层电梯厅墙面、木门窗分别涂刷为红色与白色，六层电梯厅格局变动较大，且与全楼其他部位风格迥异，略显突兀。本次修缮设计中，对于历史痕迹尚存、仅进行过涂刷改色的墙体、门窗，都按历史样式修复；而对于与历史风貌相冲突且缺少历史依据的五层电梯井红色玻璃墙和六层电梯井墙面，以能与整体协调且具可识别性的木饰面墙面及纹理玻璃进行替换。

Graceful Elevator Hall on Each Floor

In the historical drawings, the elevator halls on various floors are similar allowing for some differences. The elevator halls on the second and third floors share the same pattern, while the elevator halls on the fourth and fifth floors share the same decoration method and door/window pattern with those on other floors, but there are slight differences in the hall size and door/window details.

五楼电梯厅改造前照片
photo of the current elevator hall on the fifth floor

五楼电梯厅设计方案效果图
design sketch of the elevator hall on the fifth floor

六楼电梯厅设计方案效果图
design sketch of the elevator hall on the sixth floor

六层电梯厅格局变动较大，也与全楼其他部位历史风貌风格迥异。本次修缮设计中，拟参照历史图纸格局进行修缮，对于缺少历史依据的电梯井墙面，采用与全楼协调的"木+铜饰面"进行装饰更新，租户区入口墙面以玻璃来打造通透感。

Except for the elevator hall on the second floor whose pattern and style are almost the same as the historical drawings, the extant elevator halls on other floors have been changed. The single-leaf wooden door at the half platform of each stairwell on each floor is inferred to be the original one.

In this repair design, it is planned to restore the spatial pattern of the elevator hall on each floor with the historical pattern as a reference according to the actual situation in order to reproduce the historical features.

According to the historical drawings and mapping drawings, the spatial patterns of the elevator halls on the third and fourth floors have been slightly changed, but they are harmonious with the style of the building. Other interior decorations except for part of window openings to be adjusted according to the historical style shall be repaired based on their status quo.

In the elevator hall on the fifth floor, the walls and the wooden doors and windows are painted in red and white, respectively. The pattern of the elevator hall on the sixth floor is changed greatly, quite different from the historical style of other parts of the whole building. In this repair design, the walls, doors and windows with historical traces remaining and only painted to change colors will be repaired according to their historical styles; while the red glass wall of the elevator shaft on the fifth floor and the wall of the elevator shaft on the sixth floor which conflict with the historical features without any historical basis will be replaced with identifiable wood veneer and textured glass that can be coordinated with the whole building.

The pattern of the elevator hall on the sixth floor is changed greatly, quite different from the historical features of other parts of the building. In this repair design, it is planned to repair according to historical drawings, renew the elevator shaft wall's lack of historical basis with "wood + copper veneer" coordinated with the whole building, to create a sense of transparency with the glass entrance wall of the tenant area.

项目手记

　　城市繁华变迁，岁月痕迹留在了充满工匠精神的老建筑里。它们保存着这个城市的记忆，讲述一代又一代上海的故事。历史建筑再开发所面对的最大挑战是如何让百年建筑的生命得到延续，让新旧事物在这里碰撞交织，使历史建筑重焕活力。

　　卜内门洋行处于外滩历史风貌保护区，作为保护性建筑，临街正面不能改变，内部也要保持原样。在这个项目的改造中，深刻再现了原有建筑的精神，同时赋予了最生动的商业意识形态。而复新修葺后"新旧融合"，既传承了城市记忆建筑文化，又肩负了新的使命，匠心独具。

　　建筑改造完成后将积极引进符合项目定位和外滩金融集聚带功能定位的金融、文化、高端服务业等领域的总部型企业，以提升区域产业能级，配套服务区域整体建设大局，在此基础上打造新的中央生活圈。

　　夜幕降临，外滩万国建筑群金碧辉煌，"藏"于背后的四川中路133号大楼泛出属于自己的光芒。无论从前和现在，整栋建筑都鲜活了起来，奔流着新时代的豪情。

修缮后外立面效果图
design sketch of the facade after renovation

窗外的陆家嘴金融城
Lujiazui Financial City

Project Notes

With the changes of Shanghai, the traces of time have left in old buildings which are full of craftsmanship spirit. They keep the memory of Shanghai and tell the story for generations. The biggest challenge in the redevelopment of historical buildings is how to make their life continue, how to allow for collision of the new and old things, and how to rejuvenate them.

Brunner Mond & Co. located in the Bund Zone with Historical and Cultural Features is a protective building. Its front along the street cannot be changed, and its interior must keep intact. In the renovation of this project, the spirit of the original building has been deeply reproduced and a vivid commercial ideology has been endowed. The renovated building with a mix of old and new elements has not only inherited the culture of urban memorial architecture, but shouldered a new mission with unique ingenuity.

Upon renovation, headquarters enterprises in such industries as finance, culture, and high-end services that meet the positioning of this project and the functional positioning of the Bund Financial Agglomeration Belt will be actively introduced to improve the regional industry energy level and support the overall construction of the service area, on which basis a new central life circle will be built.

When night falls, the Bund buildings are resplendent, among which Brunner Mond & Co. gleams its unique light. The whole building is alive as before to embrace the new era.

第二章
产业复兴：苏河湾文创艺术体验空间（光复路127号）
——以新业态为导向的城市更新

Chapter Two
Industrial Revitalization: Suzhou Creek Cultural and Creative Art Experience Space
(No. 127 Guangfu Road)
– *Urban Renewal Oriented by New Business Forms*

第二章
CHAPTER TWO

　　城市化的进程仿佛一列高速行驶的子弹头列车，正在以不可阻挡的势头呼啸前进。我们看到，成群的老建筑转眼间被标新立异的高楼大厦取代，成片的老城区刹那间被游人如织的商业设施淹没。城市在生长，但文化传承的链条却出现了明显的裂痕。如果放任自流，老城的结局就是消失；如果把历史遗迹全部移进展览博物馆，老城就等于死去。

　　所谓"慢工出细活"，苏州河百联创意文化产业在社会融合、经济发展与历史遗迹保护这三者之间找到了很好的平衡点，其中最关键的一招就是——"冷水泡茶，急不得，也快不得"。

The process of urbanization is like a bullet train running at high speed, whistling forward unstoppably. As we can see, old buildings have been replaced by innovative high-rise buildings in an instant, and old urban areas have suddenly been flooded by commercial facilities. The city is growing, but there are obvious cracks in the chain of cultural heritage. With a loose rein, the old city will disappear; if moving all the historical relics into the exhibition museum, the old city will die.

As the old saying goes, soft fire makes sweet malt. Suzhou Creek Bailian creative cultural industry has found a proper balance among social integration, economic development, and protection of historical relics. Most importantly, no hurry.

屹立在苏州河畔的四行仓库和光三分库俯瞰图
aerial view of Sihang Warehouse and Guangsan Branch Warehouse on the bank of Suzhou Creek

穿越历史　邂逅经典

唤醒
RENAISSANCE

  "新"环境,"赏"新貌,"品"生活——百联资控结合"一江一河"重点景观带规划,两岸联动开发,打造苏州河百联创意文化产业轴,不仅加强了上海历史文化的底蕴,也大大提升了苏州河沿岸的地区形象,引导其从生产型岸线转变为生活型岸线,实现了真正的产业复兴。光复路127号作为苏州河百联创意文化产业轴接力的重要一棒,率先启动,将建成集商业、办公于一体的文创艺术体验空间。

  维持原样,老仓库暮气沉沉;没有修缮,新的功能无法达到应有境界。设计师充分挖掘了原有建筑的潜力,将设计难点转化为亮点,让建筑更具特色和创意。

  这座历史建筑将再次向世人敞开怀抱,以不同的风格与人文情怀融入这座城市,更完美地展现出这座国际大都市的绰约风姿。

苏州河沿岸鸟瞰图
aerial view of the Suzhou Creek

In order to "appreciate the new appearance and taste the new life in the new environment, by leveraging the development planning of the key landscape belt of Huangpu River and Suzhou Creek, Bailian Asset Holding hammers at creating Suzhou Creek Bailian creative cultural industry axis, not to strengthen the historical and cultural heritage of Shanghai, but also to greatly improve the image of the region along the Suzhou Creek and guide it to change from a production-oriented coastline to a life-oriented coastline, in which way the real industrial revitalization can be realized. As a vital part of Suzhou Creek Bailian creative cultural industry axis, No. 127 Guangfu Road will be launched first, and built into an creative art experience space integrating commerce and office.

Without any change, the old warehouses would be lifeless; without any repair, the new functions cannot play their due role. Designers are tapping into the potential of the original building to transform difficulties into highlights, making the building more distinctive and creative.

This historic building will open to the public again, and integrate into the city in a unique style and humanistic feelings to more perfectly show the graceful demeanor of this international metropolis.

溯源：时代变迁的见证

位于上海苏州河北岸，紧靠着西藏路桥的四行仓库（光复路1号）建于1931年，是座钢筋混凝土结构的六层大厦，建筑面积2万平方米，为当时闸北一带最高、最大的一座建筑物。它原是大陆银行和北四行的联合仓库。这里是人们熟知的淞沪会战四行仓库保卫战的战场，如今建成了"四行仓库抗战纪念馆"，是上海重要的一处爱国主义教育基地。

与之隔着一条晋元路的光三分库（光复路127号），是此次修缮改造项目所在。它始建于1931年，是一幢三层框架结构建筑，建筑面积4637平方米。它原是福康、福源钱庄及仓库，后为金城银行、中南银行、大陆银行及盐业银行的联合仓库，还经历了四行仓库保卫战的炮火洗礼。

回溯这座建筑经历的故事，跌宕起伏，好似一部电影，在时间的长河中，留下了一串串密码。

Sourcing: Witness to the Changes of the Times

Located on the north bank of Suzhou Creek, and near Xizang Road Bridge, Sihang Warehouse (at No.1 Guangfu Road) was built in 1931. It is a six-storey reinforced concrete structure with a building area of 20,000 square meters. It was the highest and largest building in Zhabei at that time, and used to be a warehouse shared by the Continental Bank and the other three banks. It was where the well-known Defense of Sihang Warehouse in the Battle of Shanghai took place. Sihang Warehouse has been built into Sihang Warehouse Battle Memorial, an important patriotism education base in Shanghai.

Its Guangsan Branch Warehouse (No.127 Guangfu Road), only across Jinyuan Road from Sihang Warehouse, is the project to be repaired and renovated. Built in 1931, Guangsan Branch Warehouse is a three-storey frame structure building, with a building area of 4,637 square meters. It used to be Fukang Native Bank and Fuyuan Native Bank and Warehouse, later became the warehouse shared by Jincheng Bank, Zhongnan Bank, Continental Bank and Salt Industry Bank, and survived the Defense of Sihang Warehouse.

The story of this building is like a movie, leaving a string of passwords in the river of time.

原四行仓库老照片
historical photo of the former Sihang Warehouse

原四行信托老照片
historical photo of the former Sihang Trust Department

第二章
CHAPTER TWO

苏州河对岸的四行仓库
Sihang Warehouse by the Suzhou Creek

上海自开埠以后，钱庄业逐渐兴起。太平天国战争初期，安徽籍典当业巨商程卧云到上海开办了程家在上海的第一家钱庄"延泰钱庄"。其孙程觐岳于1894年开设"福康钱庄"，福康钱庄是上海第一家做外汇业务的钱庄。1905年开设顺康钱庄。1906年延源钱庄改组为豫源钱庄，1909年，再改组为福源钱庄。

1931年，福源钱庄出资于苏州河北岸建造了三层仓库，便是如今的光复路127号——四行仓库"光三分库"。

1932年"一·二八"事变，钱庄一度停业。1933年中华民国国民政府宣布"废两改元"，1935年实行法币政策，钱庄业失去优势地位，收入大幅下降。1937年抗战全面爆发后，钱庄业务停顿。

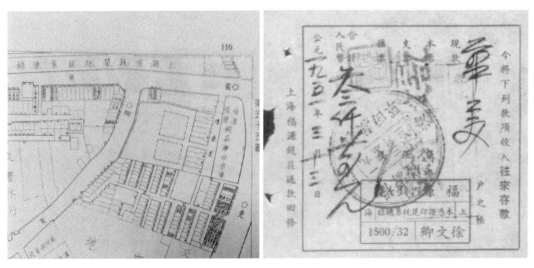

1947年版行路图
road map in 1947

1952年上海福源钱庄送款回条
payment receipt of Fuyuan Native Bank in 1952

The native banking rose after the opening of Shanghai. In the early days of the War of the Taiping Heavenly Kingdom, Anhui pawnbroker Cheng Woyun founded Yantai Native Bank, the first native bank of Cheng family in Shanghai. In 1894, his grandson Cheng Jinyue opened Fukang Native Bank, the first of its kind engaged in foreign exchanges in Shanghai. In 1905, Shunkang Native Bank was opened. Yanyuan Native Bank was reorganized into Yuyuan Native Bank in 1906 and reorganized again into Fuyuan Native Bank in 1909.

In 1931, Fuyuan Native Bank invested to build a three-storey warehouse on the north bank of Suzhou Creek, which is No.127 Guangfu Road now, the Guangsan Branch of Sihang Warehouse.

In 1932, it was shut down after the January 28 Incident. In 1933, the national government of China announced to change the currency measuring standard "Liang" to "Yuan". In 1935, due to the implementation of the Fabi policy, native banks lost the dominant position and suffered a sharp decline in income. After the outbreak of Counter-Japanese War in 1937, the business of native banks stopped.

苏州河对岸的四行仓库 ▶
Sihang Warehouse by the Suzhou Creek

溯源四行仓库（光一、光四分库）就必须要提到20世纪20年代成立的四行储蓄会。它由大陆银行、金城银行、盐业银行、中南银行这四家从中国北方起家的银行（即"北四行"），每家出资25万元成立，储户存入资金，即可保本保息，投资如有盈余，还能获得分红。这一做法一经推出，在当时即引来追捧。

1932年，四行储蓄会耗资82万元自建仓库，以堆放客户的抵押品和货物等，名曰"四行仓库"（光一、光四分库），1933年1月开始使用。1937年1月，四行信托部成立，原为交通银行的光复路195号（光二分库），原为钱庄的光复路127号（光三分库）一并移交至四行信托部管理。

四行仓库建成后，专门用于存放这几家银行客户的抵押品和货物。在当年多为二层砖木结构建筑的苏州河北岸，高达6层、建筑面积2万平方米的四行仓库宛如一座庞大的城堡。因其紧靠租界的地理位置，1937年的淞沪会战，这里成为了全球关注的焦点，八百壮士奋勇抗敌、誓死不退的事迹彪炳史册。

战后弹痕累累的"四行仓库"和西藏路桥
bullet-ridden Sihang Warehouse and Xizang Road Bridge after the war

战火中的四行仓库：硝烟右侧为四行仓库，左侧为光三分库
Sihang Warehouse in the war: Sihang Warehouse in the right and Guangsan Branch Warehouse in the left

Tracing back to Sihang Warehouse (Guangyi and Guangsi branches), we must mention Sihang Savings Society established in the 1920s. It was set up by four banks starting business from the north of China, namely Continental Bank, Jincheng Bank, Salt Industry Bank, and Zhongnan Bank (collectively called the "four banks in northern China"), with each contributing 250,000 Yuan. Depositors deposit the funds with guaranteed principal and interest. If the investment has surplus, dividends may be distributed. Once launched, this practice became popular at that time.

In 1932, Sihang Savings Society built its own warehouse at a cost of 820,000 Yuan to pile up the collaterals and goods of customers. It was called Sihang Warehouse (with Guangyi and Guangsi branches). Sihang Warehouse was put into use in January 1933. In January 1937, Sihang Trust Department was established. No. 195 Guangfu Road (Guanger Branch Warehouse), originally the Bank of Communications, and No. 127 Guangfu Road (Guangsan Branch Warehouse), a native bank at first, became in the charge of Sihang Trust Department.

Upon completion, Sihang Warehouse was specially used to store the collateral and goods of customers of the four banks. On the north bank of Suzhou Creek, where there were many two-storey brick and wood structure buildings, this six-storey warehouse with a building area of 20,000 square meters is like a huge castle. Close to the concessions, the Battle of Shanghai in 1937 made Sihang Warehouse a focus of global attention, and the story that 800 heroes fought bravely against the enemy shine throughout history.

1937年，四行孤军以步枪、机枪、手榴弹包括人肉炸弹，在此打退了以精良装备和先进战术著称的日军十余次大小猛烈进攻。尽管日军以机枪、平射炮、火攻、偷袭和毒气弹轮番上阵，但皆无功而返。

根据记载，八百壮士撤出四行仓库后，仓库曾被烈火焚烧，随后被日军侵占。抗战胜利后又一度被国民党军政机关占据，四行储蓄会费尽心力方才收回。期间还有八百壮士中的幸存者回到四行仓库，为谢晋元团长举行追悼仪式。

西墙前面的一块广场现为"晋元纪念广场"
Jinyuan Memorial Square in front of the west wall

20世纪90年代后期四行仓库被用作为上工批文化礼品市场
Sihang Warehouse used as Shanghai Industrial Wholesale Cultural Gift Market in the late 1990s

1949年后，这里依旧长期被用作仓库，1985年上海市文保委在此勒石纪念，上书"八百壮士四行仓库抗日纪念地"；1993年后，四行仓库先后成为上海工业品批发市场、上工批文化礼品市场。2015年，光复路127号光三分库被列入上海市第五批优秀历史建筑之列。

20世纪30年代 1930s　　　　20世纪90年代 1990s　　　　现在 present

In 1937, with rifles, machine guns, grenades, and even human bombs, these warriors fought back more than ten large and small fierce attacks by the Japanese army famous for its excellent equipment and advanced tactics. The Japanese attacks with machine guns, flat fire guns, fire attacks, sneak attacks and poison gas bombs all failed.
According to records, after 800 warriors withdrew from Sihang Warehouse, the warehouse was burned by fire and then occupied by the Japanese army. After the victory of the Counter-Japanese War, it was once occupied by the military and political organ of the Kuomintang, and later recovered by Sihang Savings Society with great efforts. During this period, the survivors of the 800 warriors returned to Sihang Warehouse to hold a memorial ceremony for Commander Xie Jinyuan.
After 1949, it was still used as a warehouse for a long time. In 1985, Shanghai Municipal Committee of Heritage Protection erected a stone tablet which writes "Memorial to 800 Heroes against the Japanese Army in Sihang Warehouse"; after 1993, Sihang Warehouse successively became Shanghai industrial wholesale market and Shanghai industrial wholesale cultural gift market. In 2015, Guangsan Branch Warehouse at No. 127 Guangfu Road was listed among the fifth batch of outstanding historical buildings in Shanghai.

循迹：于空间中触摸建筑的温度

从苏州河南岸望去，可以完整地看到四行仓库东西两楼概貌。仓库的西面墙体弹孔累累，上面两个当年被日军平射炮击穿的墙洞赫然再现。经过枪击炮轰后，整个西墙黝黑、破败不堪。进入仓库后，压抑感在阴暗的光线中扑面而来。当年，谢晋元在仓库里对士兵们说："四行仓库就是我们最后的阵地，就是我们的坟墓。"

四行仓库西面墙面现状照片
photo of the current west wall of Sihang Warehouse

四行仓库现状照片
photo of the statuo quo Sihang Warehouse

跨过窄窄的晋元路，就是光三分库，红墙白立柱将整座建筑的外立面装饰得分外靓丽，与另一侧灰白色调的四行仓库对比鲜明。

仓库外立面为现代主义建筑风格，局部做简洁的 **ArtDeco** 装饰。红墙白柱，柱间是黑色油漆面的横向钢质长条型高窗，设置有精致的钢质五金件，窗的分隔充满韵律感，具有典型仓储建筑特征。

Tracking: Touch the Temperature of the Building in the Space

Looking from the south bank of Suzhou Creek, you can see the complete profile of the east and west buildings of Sihang Warehouse. There are two holes pierced by Japanese flat fire artillery on the bullet-ridden west wall. After being shot and shelled, the west wall is dark and dilapidated. Entering the warehouse, a feeling of depression overwhelms amidst the dark light. At that time, Xie Jinyuan said to the soldiers in the warehouse, "Sihang Warehouse is our last position and our grave."

Guangsan Branch Warehouse stands across the narrow Jinyuan Road. The red wall and white columns highlight the well-decorated facade, which is in contrast with Sihang Warehouse in gray and white on the other side.

The facade is of modernist architectural style, decorated with simple Art Deco partially. It features red wall and white columns, between which are horizontal steel long-strip high windows with black finishing paint set with exquisite steel hardware. The separation of windows is full of rhythm, with characteristics of a typical storage building.

光复路127号外立面现状图
photo of the appearance of current No. 127 Guangfu Road

仓库原设计在南面设置主出入口、楼梯间通廊作主要交通空间，两侧分层设置仓储空间，南北向连通，布局高效科学；同时，针对交通和仓储不同的使用要求，采用梁柱框架结构体系，反映了现代建筑功能主义特点。

建筑平面大致呈梯形，南立面为弧形，房屋中部偏西处设南北向伸缩缝，将房屋分为东、西两个结构单元。房屋外墙及内部墙体主要采用烧结普通砖砌筑。宽敞平缓的混凝土楼梯具有仓储建筑特色，底层门厅为水泥地坪。

In the original design, the main passageway on the south facade and the corridor of stairwells are set as the main traffic space, while the storage space is set in layers on both sides to connect the north and the south, which is efficient and scientific. Besides, according to the different use requirements of traffic and storage, a beam-column frame structure system is adopted, which reflects the features of modern architectural functionalism.

The building plane is roughly trapezoid. The south facade is arc-shaped. A north-south expansion joint is set in the west of the middle part to divide the building into two structural units, the East and the West. External and internal walls are mainly made of sintered common bricks. The spacious concrete stairs have the characteristics of a storage building, and the lobby on the ground floor is paved with cement.

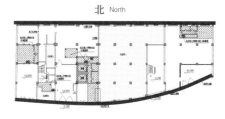

光三分库平面图
plan of Guangsan Branch Warehouse

楼梯间现状照片
photo of the current stairwell

根据1947年版行号图及现场查勘分析，建筑分期建设，东侧七跨建设较早，西侧四跨建设较晚，整体竣工时间相差不远。根据不同时期的图纸资料发现，该建筑于1967年、1990年、1992年经过多次改造。1967年根据厂库使用需求，在内部增加两台电梯；1990年在四层加建办公楼及电梯机房225平方米；1992年在四层西侧加建171平方米。

仓库南面入口处钢门已经损毁，部分窗洞封堵，部分窗已更换，硕大的"四行仓库"金属字及防盗网均为后期增加，并非历史原物；墙面及窗台、勒脚等部位被上次修缮时的涂料饰面覆盖，架空线路如蛛网错综复杂，严重影响了建筑立面的完整性。

According to the commercial map of 1947 and on-site survey and analysis, the building was constructed by stages. The seven bays in the east were constructed earlier than the four bays in the west, but the overall completion time is not far from each other. According to the drawings in different periods, this building was renovated several times in 1967, 1990 and 1992. In 1967, to satisfy the use demand of the factory warehouse, another two elevators were built inside; in 1990, an office building and an elevator machine room covering an area of 225 square meters were added on the fourth floor; in 1992, 171 square meters were added in the west of the fourth floor.

The steel door at the south entrance of the warehouse has been damaged, some window openings have been sealed, and some windows have been replaced. The large metal characters "Sihang Warehouse" and the burglar mesh are both added later, not the historical original ones. The parts including walls, windowsills, and plinths are covered by the finishing paint in the last repair. Overhead lines are as complex as a cobweb, seriously affecting the integrity of the building facade.

仓库南面现状照片
photo of the current south facade of Sihang Warehouse

光复路127号修复前鸟瞰图
aerial view of No. 127 Guangfu Road before restoration

  走进仓库内,岁月加之于它的沧桑印迹亦随处可见——楼梯水泥踏步肉眼可见的磨损,楼梯扶手锈蚀、部分损毁,墙面污损、粉刷脱落的情况无处不在。几扇黑漆铁门显得庞大而厚重,油漆脱落锈蚀,就如同那段湮没在时光长河里的悲恸历史。空荡荡的仓库里,唯有一根根柱子依旧挺立,部分梁柱风化开裂,墙皮脱落。柱子、墙壁上仍留有那个年代的红色标语,向人们提示着它的过去。

程大利女士在光复路127号现场考察
on-site survey of No. 127 Guangfu Road by Ms. Cheng Dali

爬上顶层，视线豁然开朗。南面的苏州河宛如一条流动的绿色丝带，对岸林立的高楼勾勒着曲折生动的天际线，楼下临河绿化景观带设置了舒适宜人的步道，是休闲散步的好去处。

　　顶层加建的房屋显得突兀生硬，散落的绿植盆栽，散发出顽强的生命力，仿佛也在寓示着这栋老仓库渴望迎来新生机。

　　根据建筑专家结构检测，整栋仓库大楼主体结构目前存在一定结构性损伤，部分楼层的楼板承载力普遍不满足计算要求。整栋建筑仿佛进入了残朽暮年，亟待一场"重新活过来"的保护性修缮改造。

Entering the warehouse, the traces of time are everywhere: visibly worn cement stair steps, corroded and partially damaged stair handrails, stained walls, and fallen paint. There are several huge and heavy black painted iron doors with fallen and rusted paint, just like the mourning history buried in the river of time. In this empty warehouse, only one column stands still, some beam columns are weathered and cracked, and the walls have peeled off. There are also revolutionary slogans of that era on the columns and walls to remind people of its past.

Climb up to the top floor, and you'll embrace an open and clear vision. Suzhou Creek in the south is like a flowing green ribbon. The tall buildings on the opposite bank outline the twisted vivid skyline. The green landscape belt by the creek creates a comfortable and pleasant footpath for leisure and walking.

The house added on the top floor is abrupt and stiff. The scattered green plants give out strong power of life, as if they also indicate that the old warehouse is eager to usher in new vitality.

According to the structural inspection of architectural experts, the main structure of the warehouse building is subjected to some structural damages, and the bearing capacity of some floors is generally below the calculation requirements. The building seems to have entered its old age, in an urgent need of a protective renovation for resurrection.

斑驳的窗外树木郁郁葱葱　　　　　　仓库内部现状照片
luxuriant trees outside the mottled windows　　　photo of interior of the current warehouse

更迭：为建筑遗产吹入生命的气息

当下，建筑不再过多地执着于成为城市天际线上的地标，转而开始关注建筑保护、建筑与城市的有机和谐，将更多关注的焦点放在建筑的历史文化、环境、功能、空间、技术、材料等方面，这是从地标主义向"传承"的根本性转变。

挺过了炮火硝烟，穿过了计划经济、市场经济的浪潮更迭，时间进入2019年，光复路127号光三分库正是在这种背景下进行的一次意义重大的整修与翻新。

作为上海市人民政府批准公布的上海市第五批优秀历史建筑（保护类别三类），百联资控整体遵循上海市历史建筑保护事务中心的修缮意见，在权威专家的指导下，尊重建筑本体的历史价值，推陈更新，赋予四行仓库"光三分库"全新的生命力。

根据《关于光复路127号优秀历史建筑修缮工程保护要求告知单》内容，房屋南、北立面为外部重点保护部位，门厅、楼梯间、电梯、各层黑漆仓库门、金属门窗构件以及其他原有特色装饰等为内部重点保护部位进行全面修缮。根据保护要求和项目使用功能情况，对室内公共部位进行装修。20世纪90年代四层加建部分，根据保护要求及现场情况局部调整，拆除部分加建面积。

Changing: Inject Life into Heritage Buildings

Buildings have no longer attached much to becoming a landmark on the urban skyline, but to the protection and their organic harmony with the city. More attention has been paid to such aspects of buildings as historical culture, environment, functions, space, technology, and materials, marking a fundamental change from landmark orientation to inheritance orientation.

Having survived the gunfire and experienced the change from planned economy to market economy, in 2019, Guangsan Branch Warehouse embraced a renovation and renewal of great significance.

Listed in the fifth batch of outstanding historical buildings in Shanghai (under grade III protection) announced by Shanghai Municipal People's Government, Bailian Asset Holding will renovate Guangsan Branch of Sihang Warehouse and inject new vitality on the premise of following the repair opinions of Shanghai Historical Building Protection Affairs Center, and respecting the historical value of building noumenon under the guidance of authoritative experts.

According to the *Notice on Protection Requirements for Repair Engineering of Outstanding Historical Building No. 127 Guangfu Road*, in this renovation, the south and north facades are key exterior protected parts, while the lobbies, stairwells, elevators, black paint warehouse doors, metal door/window components and other original characteristic decorations are key interior protected parts. According to the protection requirements and use functions, the indoor public parts shall be decorated. The space added on the fourth floor in the 1990s shall be partially adjusted according to the protection requirements and site conditions, and part of which shall be demolished.

整个项目改造遵循"保护为主、合理运用、加强管理"的方针,正确处理保护和合理利用关系,力争最大限度保留建筑本体原物、原件以及其本身负载的历史、文化、社会的各方面有价值信息,使建筑的原真性得到最大程度的保持,并清晰可读。在保护历史建筑的原真性与完整性的前提下,再充分考虑未来现代综合性商业、商务建筑的各种功能要求。

经过修缮改造后的光三分库,将被打造成一张苏州河畔文创艺术体验空间的亮眼名片,引入艺术书展、讲座、市集等各种文化艺术交流活动,为时尚创意产业打造集创作、展示、体验与销售为一体的创新生态。二层至三层为标准办公空间,吸纳创意设计类企业和高端办公服务类的企业入驻,首层则以配套精品商业服务办公企业的需求。顶层露台打造景观酒吧餐厅,希望未来会成为追求时尚体验的年轻人竞相打卡的城市网红聚集地。

In line with the principle of "protection first coupled with rational use and management", this renovation will properly handle the relationship between protection and rational use to maximally keep the original items of the building and its historical, cultural and social valuable information, thus maximizing the authenticity of the building and making it clear and readable. The future functional requirements for modern comprehensive commercial and business buildings should be considered on the premise of protecting the authenticity and integrity of historical buildings.

The renovated Guangsan Branch Warehouse will be built into an impressive Suzhou Creek cultural and creative art experience space that renders a creative ecosystem integrating creation, display, experience and sales for the fashion and creative industry by introducing such cultural and artistic exchange activities as art book exhibitions, lectures, and market fairs. The second to third floors will be equipped with standard offices to attract creative design enterprises and high-end office service enterprises. In contrast, the ground floor will meet the need of supporting high-quality commercial service office enterprises. The landscape bar-restaurant on the top floor is expected to be an internet-famous site for young people who seek fashion experience in Shanghai.

领导专家在光复路127号现场考察
On-site survey of No.127 Guangfu Road by leader and expert

唤醒
RENAISSANCE

光复路127号改造日景效果图
daytime renovation design sketch of No. 127 Guangfu Road

鼎新：在新旧平衡中找到独特建筑语言

  本次改造并非重新建造一个全新的立面，而是从仓库的原始功能出发，找到属于建筑自身特质的独特语言，并将其运用到改造的设计中去。

  空间在新与旧的冲击下，尽力以保持克制来达成最持久的视觉张力，新旧元素结合延续了建筑的记忆，又为这里的工作和娱乐休闲提供了一种全新的体验。

Innovating: Find a Unique Building Language in the Balance between New and Old Elements

This renovation is not to rebuild a new facade, but to find a unique language that tallies with the characteristics of the warehouse from its original function and apply it to the renovation design.
The space tries to achieve the most lasting visual tension with restraint, so that the mix of new and old elements may continue the memory of the building to offer new experience for work and entertainment here.

光复路127号改造夜景效果图　nighttime renovation design sketch of No. 127 Guangfu Road

建筑将实施必要的结构加固和给排水、电气、暖通工程修缮,然后对外墙进行杂物清理、清洗修复,南墙上的"四行仓库"金属字也将被拆除。

首层南立面除大铁门外均为卷帘门店铺,门窗在后期改建过程中变得非常凌乱,尺寸形式不一。遵循最小干预原则,对已开洞的门面统一修缮为商业门面。将南立面首层高窗按历史风貌恢复;按原仓库墙面和高窗比例设置下部商业门面;商业门面延续了上部钢窗的形式,设计为铝合金仿钢材玻璃门,做到与立面整体风格协调。

The necessary structural reinforcement and repair of water supply and drainage, electrical, heating and ventilation works will be done. The exterior wall will be rinsed and repaired, and sundries on which will be cleaned. The metal characters of Sihang Warehouse on the south wall will also be removed.

In the south facade of the ground floor, except for the large iron gate, other premises are rolling shutters. The door windows are rather messy in different sizes and forms after several renovations. In line with the principle of minimal intervention, the doors with holes will be uniformly repaired into commercial facades. The high windows on the south facade of the ground floor will be restored according to their historical features. The lower commercial facades will be set according to the proportion of the original warehouse wall and high windows. The commercial facades will continue the pattern of the upper steel windows, and use aluminum alloy quasi-steel glass doors to coordinate with the overall style of the facade of the whole building.

光复路127号现存墙面
existing walls of No. 127 Guangfu Road

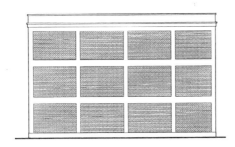
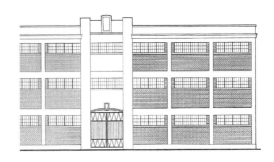

光复路127号工程设计图纸
design drawing of No. 127 Guangfu Road project

楼梯间现大门非历史原物，将参照同期仓库大门样式恢复。四层原加建部分南侧靠近外墙一跨拆除，西侧墙体拆除，保留加建的框架柱，加建部分高度参差不齐，整体统一修缮。

清水砖外墙面按照"清理-清洁-旧砖置换与修补-砖缝整修-表面防渗保护"的工艺流程严格修缮。外墙的装饰水泥线条、水泥砂浆墙柱面也将得到清洗修补。本项目由于室内地坪整体抬高，原底层防潮层已全部失效，所有防潮层失效的砖墙均需在墙体底部增设新的防潮层。

作为外墙改造亮点的东立面，为非重点保护部位，根据使用功能开消防专用门窗，在保留原框架体系的基础上，填充墙部位改为丝网印刷红砖效果玻璃，既保留原仓库建筑特点，又改善室内采光条件，也可遥望对面的四行仓库抗战纪念馆，与历史隔空对话。

东立面现状和改造效果图对比
Comparison of current photo and renovation design sketch of the south facade

室内修缮范围和内容包括门厅、楼梯间、电梯厅、二三层公共走道、各层黑漆仓库门、金属门窗构件以及其他原有特色装饰等，底层还将新增设备用房。所有钢窗、钢门都将进行整体检修，并刷暗黑色油漆以恢复原貌，做到美观和适用性兼具。

The extant gate of the stairwell is not the original one. It will be restored in reference to the warehouse gate pattern in the same period. The additional bay in the south near the exterior wall on the fourth floor will be removed together with the west wall, but the additional frame column will be retained. The additional part is of various heights, so it will be repaired wholly.

The exterior fair brick wall will be repaired strictly according to the process of "clearing–cleaning–replacement and repair of old bricks–repair of brickwork joints–surface anti-seepage protection". The decorative cement lines and the cement mortar cylindrical surface of the exterior wall will be cleaned and repaired. In this project, due to the overall elevation of the indoor floor, the original damp proof course of the bottom floor has lost efficacy. All brick walls with a failed damp proof course shall be added with a new one.

The east facade, a highlight in exterior wall renovation, is not a key protected part. According to the use function, fire protection door windows will be added. Based on retaining the original frame system, the infilled wall part will be changed into silk-screen printing red-brick effect glass to not only retain the features of the original building as a warehouse, but improve the indoor lighting conditions and echo with Sihang Warehouse Defense Memorial in the opposite.

Indoor repair involves the lobby, stairwells, elevator halls, public walkways on the second and third floors, black paint warehouse doors on each floor, metal components on door windows and other original characteristic decorations, and a new equipment room to be added on the ground floor. All steel windows and doors will be overhauled, and painted with dark black paint to restore the original appearance for both beauty and applicability.

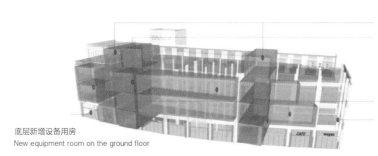

修缮重点保护部位楼梯间
Stairwells as a key protective part

一至四楼电梯厅装修
Elevator halls from the ground to fourth floors

二、三层公共走道装修
Public walkways of the second and third floors

底层新增隔油池间
Separation tank room to be added on the ground floor

底层新增设备用房
New equipment room on the ground floor

室内改造示意图
Design sketch of indoor renovation

南面楼梯间立面恢复洞口及历史栏杆。但设计师不主张在楼梯上增加其他装饰性物件，坚持使其回复到历史原貌状态，仅仅在原栏杆内增加一道0.9米高的玻璃栏板防护，满足安全要求；并在原内墙墙面增加楼层指示标志。

In the south stairwell, the opening and historical handrail will be restored. However, the designers do not advocate to add other decorative objects on the stairs, but to return to the original state, only adding a 0.9-meter-high glass fence protection inside the original handrail to meet the demand of safety; and a floor indication sign will be added on the original interior wall.

楼梯间改造效果图
Renovation design sketch of the stairwell

电梯厅及入口接待台改造效果图
Renovation design sketch of the elevator hall and the reception desk at entrance

电梯厅采用灰色水磨石地坪，入口接待台采用同材质。原白色墙面进行修缮，局部以清水水泥墙面及艺术画装饰，加之灰色金属网格吊顶，使电梯厅既保留原有的仓库建筑风格又具有现代商业办公建筑特色。

The elevator hall will adopt gray terrazzo floor, so will the reception desk at entrance. The original white wall will be repaired, partially decorated with fair cement wall and art paintings, together with a gray metal grid ceiling to make the elevator hall feature both original warehouse-style and elements of a modern commercial office building.

公共区域改造效果图
renovation sketch of the public area

唤醒
RENAISSANCE

保持旧，也创造新。旧得韵味犹存，新得现代时尚。留旧向新的转变，是一种超越。

旧——室内特色的大铁门会保留下来，清洗除锈，但不再具有实际功能，仅作为历史符号存在，起装饰作用。

新——顶层露台发挥其景观优势，改造成城市露台酒吧和餐饮。

黑漆铁门将作为空间的历史符号
Black paint iron gate as a historical symbol of space

We are dedicated to innovation while retaining the essence of the old style for the coexistence of classical charm and modern fashion. Changing from the old to the new is also a kind of transcendence.
Old—the unique large indoor iron gate will be retained, cleaned and derusted, without any practical function, only for decoration as a historical symbol.
New—the top floor will play its landscape superiority and be renovated into an open bar and catering space.

景观餐厅示意图
Design sketch of the landscape restaurant

　　光复路127号的改造，以创意、文创艺术体验空间为核心，兼备社会效应和市场效应。

　　不同于CBD鳞次栉比的办公大楼给人压抑、紧迫、拘谨的氛围，在此，既能享受城市核心地段的办公便利，又能体会老建筑开放、自由和开阔的胸怀，更能在苏州河这一传承着上海浓厚历史底蕴的文脉之地，悦享一段绝美的临河风光和城市休闲景观露台，于工作于生活，恣意舒展年轻的灵魂。

The renovation of No. 127 Guangfu Road takes creativity and cultural and creative art experience space as the core, with both social and market effects.
Unlike the lining oppressive and restrained office buildings in CBD, this building, which features rich historical heritage of Shanghai and is perfect for work and living, enables you to not only enjoy the office convenience in the core area of Shanghai, but experience the openness, free and broadness of an old building, and enjoy a beautiful riverside scenery and urban leisure landscape floor by Suzhou Creek.

项目改造后效果图 penovation design sketch of the project

第二章 CHAPTER TWO

项目手记

当修缮从方案草图落于实际施工过程中，我们会发现历史建筑所历经的沧桑远超越我们最初的认知。在光复路127号，多处外墙不规则开裂、窗台破损、二三层地坪不平，而且由于东西两侧建筑相隔两年建成，两侧地坪高度也存在不统一的情况，甚至建筑还存在着严重沉降，高差达到了9厘米。这将大大增加修缮施工的难度，并要求建筑师们在进场后针对性地提出各类抢修方案。

而这，也正是历史建筑的价值所在——旧的建筑躯壳已然腐朽，新的骨骼亟待被修正构建，才会在时代的潮流中激荡出愈加精彩的空间功能，让老仓库在苏州河沿岸找到自己傲然独立的位置，继续讲述它们精彩的故事。

当我们穿越历史俯仰在过去与未来之时，老仓库背后的故事已被收入纪念馆藏，它的未来更值得我们期待。百联资控将践行"以人为本"的理念，重塑一个全新的光三仓库；而城市更新的全新格局，也将在这里风云际会。让我们共同期待着这座充满沧桑的历史建筑在黯然数十年后，再一次惊艳苏州河畔，迎来它的华丽转身。

Project Notes

In the process from the scheme sketch to the actual construction, we will find that the vicissitudes a historical building has suffered go far beyond our initial cognition. At No. 127 Guangfu Road, there are many irregular cracks on the outer wall, damaged windowsills, uneven second and third floors, disunited floor heights in the east and the west due to the construction interval of two years, and even the serious subsidence deep up to 9 cm, which will greatly increase the difficulty of repair construction, and require architects to put forward various emergency repair plans after entering the site.

And this is the value of a historical building—its old body has been decayed, a new framework needs to be revised and constructed urgently to render more fantastic space functions in the new era, so that this old warehouse can find its unique position by the Suzhou Creek, and continue to tell its wonderful stories.

When we envisage the past and the future through history, the stories behind the old warehouse have been collected in the memorial hall, but its future is more worthy of our expectation. Bailian Asset Holding will practice the people-oriented concept to reshape a brand-new Guangsan Branch Warehouse; and the new pattern of urban renewal will also take shape soon. Let us expect this time-honored historic building to once again amaze people by the Suzhou Creek after silence for decades in this new era.

唤醒
RENAISSANCE

第三章

创新驱动：百空间（天津路50-66号）
—— 身在"故事"、创造"故事"

Chapter Three

Innovation Driving: 百空间 (No. 50-66 Tianjin Road)

– Create a New "Story" in the Old "Story"

第三章 CHAPTER THREE

建筑更新在我国早已成为了一种趋势，在以功能置换等方式进行的老建筑再利用的开发中，**百空间**系列产品走在了前端并引领潮流。

Building renewal has already become a trend in China. In the development of the reuse of old buildings by such means as function replacement, **百空间** series products spearhead and lead the trend.

天津路50-66号 **百空间** 实景图
view of **百空间**

唤醒 RENAISSANCE

天津路 百空间 实景图
photo of 百空间

2018年，百联资控与全球领先的创造者社区、空间和服务提供商WeWork在上海正式对外宣布达成深度合作，共同参于激活城市历史建筑及存量建筑，进一步盘活优质资产，提高经济效益，为更多区域带来社区活力，助力城市复兴，为更多本地创造者提供全新工作及生活方式。

百联资控希望通过合作能有效提升存量资产的市场价值，整合百联集团资源，激活城市活力，使百联在上海城市更新的进程中迈出创新的一步。

首个改建落成的 百空间 位于上海市黄浦区天津路50—66号，地理位置优越，距离外滩780米、南京东路步行街190米，邻近地铁2号线、10号线南京东路站，周边汇集上海新世界大丸百货、宏伊广场、恒基名人购物中心等商业项目，项目建筑面积约3200平方米。

天津路 百空间 实景图
photo of 百空间

In 2018, Bailian Asset Holding and WeWork, a global leading creator community as well as a space and service provider, officially announced in Shanghai that they had reached in-depth cooperation in activating urban historical and in-stock buildings to further invigorate high-quality assets, improve economic benefits, bring community vitality to more regions, help the city to revive, and provide new work and life ways for more local creators.

Bailian Asset Holding hopes that through cooperation, the market value of in-stock assets can be effectively enhanced to activate the vitality of Shanghai by pooling the resources of Bailian Group, and take an innovative step amid urban renewal.

The first completed renovation project 百空间 which is located at No. 50-66 Tianjin Road, Huangpu District, Shanghai. It enjoys location superiority: 780 meters away from the Bund, 190 meters away from Pedestrian Street on East Nanjing Road, and near the East Nanjing Road Station of Metro Line 2 and Line 10. There are many commercial projects including Shanghai New World Daimaru Department Store, Hongyi Plaza, and Henderson Metropolitan as well. The construction area of this project is ~3,200 square meters.

天津路项目旨在使用当代设计元素创造出一个独特的高品质办公空间。旧时的钱庄银行摇身一变成了21世纪的时髦办公空间。在这个充满生命活力的魔力大都市，传统和现代思维碰撞，复古和时尚灵魂交错。历史保护建筑天津路66号记录着时光飞逝中历史文脉的沉淀，也见证着城市飞速发展的日新月异，在精心打造下，它愈加散发出振奋人心的激情和活力。

This project aims to create a unique high-quality office space using contemporary design elements. The old native bank has become a fashionable office space in the 21st century. In this magic metropolis which is full of vitality, traditional and modern thinking collide; retro and fashion elements interlace. The protective historical building of No. 66 Tianjin Road records the precipitation of historical context in the fleeting time, and witnesses the rapid development of Shanghai. Through elaborate renovation, it is sending forth greater passion and vitality.

天津路66号门头实景图
photo of the front of No. 66 Tianjin Road

百空间 正门入口实景图
photo of the front entrance of 百空间

演绎旧上海金融往事

天津路是位于上海市黄浦区的一条街道，东西走向，东起江西中路，西至贵州路。宽2米到12米，长1045米，不过是公交线路两个车站的路程。最初由上海英租界当局修筑于1851年，因通往抛球场，名为球场弄或五柱球弄，俗称"后马路"（对应于南京路即"大马路"）。1865年，上海公共租界工部局正式定名为天津路，得名于新开辟的通商口岸天津。

天津路为上海市中心区的传统商住混合区，两侧多石库门里弄住宅，1949年以前分布有许多银行、钱庄，例如2号（江西路口）的广东银行（今光大银行）、100号（河南路口）的恒利银行（今永利大楼）。

天津路一段街景
streetscape on Tianjin Road

天津路上光大银行照片图
photo of Everbright Bank on Tianjin Road

Recurrence of the Financial Story in Old Shanghai

Tianjin Road lies in Huangpu District, Shanghai, stretching from Middle Jiangxi Road in the east to Guizhou Road in the west. It is 2-12 meters wide and 1,045 meters long, only the distance of two bus stops. Originally built by the British concession authority in Shanghai in 1851, Tianjin Road was called Park Lane or Five-post Park Lane due to its access to the throwing court, commonly known as the Back Road (relative to Nanjing Road, i.e., the Main Road). In 1865, it was officially named Tianjin Road by the Shanghai Municipal Council after the newly opened trading port Tianjin.

天津路50-66号建筑特写
close-up photo of No. 50-66 Tianjin Road

Tianjin Road is a traditional area intermixing commerce and residence in downtown Shanghai, with many Shikumen houses on both sides. Before 1949, there were many native banks including Guangdong Bank (now Everbright Bank) at No.2 (at the intersection of Jiangxi Road) and Mercantile Bank (now Wing Lee Mansion) at No.100 (at the intersection of Henan Road).

天津路不远的外滩手绘图
freehand sketching of the Bund near Tianjin Road

短短的天津路蕴藏着一段旧上海的金融往事。

天津路是一条与南京东路平行相隔的横马路，但它可没有捞到什么"二马路"的头衔，它的名气不靠南京路，而是来自这一条路上耀眼的钱庄。那时的上海，钱庄最集中的地方就是天津路，一直到1949年，全市还有钱庄80家，其中天津路和宁波路联袂占据了半壁江山，达到了40多家，是老上海名副其实的"钱庄街"。

上海有钱庄的历史是在1776年前。据《上海钱庄史料》记载，上海最早的钱庄是由一个绍兴商人所设的煤炭店转而"专以兑换银钱及放款于店铺船帮，逐渐扩大，形成此后之钱业。"

自1843年上海开埠以来，钱庄的重心就逐渐从南市的豫园移至北市的租界；太平军进攻上海，尤其是1853年（即清咸丰三年）的小刀会又加速了这个转移的进程，战乱使钱庄和资本纷纷移入租界避难，后来都集中到与外滩银行相邻的天津路和宁波路，成为新的钱庄街。

这段旧时金融圈的过往可谓圈定了天津路在当时上海金融圈的地位，天津路上的钱庄也可以说是多到近乎是比肩接踵的程度。不仅沿路的钱庄一家挨着一家，就连马路沿线的一些弄堂里也开着不少的钱庄，有福绥里、同吉里、鸿仁里……当时上海最大的钱庄——福康钱庄和顺康钱庄也落户于此，一番繁荣兴旺的景象。

On the short Tianjin Road, there was a history of finance in old Shanghai.
Tianjin Road is a cross road parallel to East Nanjing Road, but it did not get the title of "the Secondary Road". Its fame did not due to Nanjing Road, but from the famous native banks on it. At that time, Tianjin Road gathered the most native banks in Shanghai. In 1949, there were about 80 native banks in Shanghai, over 40 of which were on Tianjin Road and Ningbo Road. It is a real street of native banks in the old Shanghai.

一份旧地图上标注了当时的
天津路50号为中汇银行，66号为惠中银行
No. 50 Tianjin Road (New Chung Wei Bank) and
No. 66 Tianjin Road (Huizhong Bank)
marked in an old map

新旧建筑的隔空对话
dialogue between old and new Buildings across time

The history of native banks in Shanghai can be traced back to around 1776. According to the *Historical Materials on Native Banks in Shanghai*, originally, a Shaoxing businessman "began to exchange money and lend to shop and ship operators in his coal store, gradually, the native banking was formed."

After the opening of Shanghai in 1843, the focus of native banks in Shanghai was gradually shifted from Yu Garden in Nanshi to the concession in Beishi. Due to the attack of Taiping Army on Shanghai, coupled with the Small Sword Society in 1853 (the third year of Xianfeng in the Qing Dynasty), both native banks and capital moved to the concession. Later, they concentrated on Tianjin Road and Ningbo Road, near the Bund banks at that time, which became the new street of native banks.

This history of finance defined the position of Tianjin Road in the financial circle of Shanghai at that time, and the native banks on Tianjin Road were quite close to each other. There were many native banks even in some alleys including Fusui Alley, Tongji Alley and Hongren Alley along the Tianjin Road. Fukang Native Bank and Shunkang Native Bank, the largest ones in Shanghai at that time, were also settled here, contributing to the prosperous scene.

天津路50—66号建筑实景照
Photo of the building at No. 50-66 Tianjin Road

  本项目的主角天津路50—66号建造于1934年，是坐北朝南的三层钢筋混凝土结构，建筑面积为2546平方米。中汇银行与惠中银行曾设于此，这里也曾经是上海市银行旧址，1938年后因战事原因停业，抗日战争胜利后，银行复业，1949年5月27日被上海市军事管制委员会接管。1958年6月起由房管部门管理，后由上海市燃料公司等单位使用。

Built in 1934, No. 50-66 Tianjin Road is a three-storey reinforced concrete structure facing south, with a building area of 2,546 square meters. New Chung Wei Bank and Huizhong Bank were once located here, so did the Bank of Shanghai, which was closed down after 1938 because of the war. After the victory of the Counter-Japanese War, the bank resumed its business and on May 27, 1949, it was taken over by the Shanghai Military Control Committee. It was managed by the Shanghai Municipal Bureau of Housing Security and Management from June 1958 and then used by Shanghai Fuel Company and other entities.

光影记录着时光飞逝里历史文脉的沉淀
也见证着城市飞速发展的日新月异
The light and shadow record the precipitation of historical context in the fleeting time, and witness the rapid development of Shanghai.

老建筑独有的历史印记
historical imprint unique to the old building

　　这座曾经被上海市银行，惠中商业储蓄银行、惠中产物保险公司、国华烟厂股份有限公司发行所先后设立办公场所的大楼，感受着这座城市的变迁、发展，见证着它的日新月异。一栋楼，一个走廊，一砖一瓦，承载的都是满满的回忆。

　　如今的天津路已经发生了极大的变化，马路拓宽、山东北路接通、旧城改造等，拆除了许多旧建筑，周围如雨后春笋般涌现各式各样的商场、医院、商务楼。

This building, which was used as a premise by the Bank of Shanghai, Huizhong Commercial Savings Bank, Huizhong Property Insurance Company, and Distribution Agency of Kuo Hua Tobacco Factory Co., Ltd., has experienced and witnessed the changes and development of Shanghai. It is full of memories.
Today's Tianjin Road has undergone great changes. The road has been widened and connected to North Shandong Road. The old urban area has been renovated. Many old buildings have been demolished. Various shopping malls, hospitals and business buildings have sprung up.

天津路50—66号项目俯瞰，对面就是赫赫有名的新世界大丸百货大楼
aerial view of No. 50-66 Tianjin Road, with the famous New World Daimaru Department Store of the street

沿袭书写历史文脉

老建筑奏响全新乐章，走进焕然一新的天津路百空间，改造团队仍能勾起改造之初所面临的种种困难的回忆。

天津路50—66号系黄浦区文物保护点，修缮这座建筑时，需要遵守不改变文物原状的原则，对建筑主立面的修缮也要结合历史资料等进行考证，尽可能在保存其真实历史风貌的同时，进行架构修缮加固，增加安全性。例如外墙靠近窗台处别致的卷叶雕花，透露着旧时的记忆。这是老建筑自有的特色细节，势必要仔细修缮，放大其华彩。

Inheritance of Historical Context

别致的卷叶雕花
透露着旧时的记忆

外墙重新涂刷，卷叶雕花也得以精心修缮
Repainted exterior wall and carefully repaired curly leaf carving

The old building has had a new appearance. Entering the completely new space of 百空间 at Tianjin Road, the renovation team can still recall the difficulties they faced at the beginning of renovation.

No. 50-66 Tianjin Road is a cultural relic protection site in Huangpu District. During the repair, the principle of not changing the original state of the cultural relic was followed. The repair of the main facade has also been based on surveys such as historical data for structural repair and reinforcement to increase safety while maximally preserving its true historical features. For example, the unique curly leaf carving, which is near the window sill and on the facade, reveals the memory of the past, so it must be carefully repaired to magnify its beauty.

悬挂着文明单位的天津路50—66号 ▶
No.50-66 Tianjin Road with model unit nameplate

项目在1934年建造之时是三层，建筑面积为2546平方米，现存的第四层是20世纪80年代加盖，因此加层建筑与老建筑之间的构造关系也需要梳理清楚，仔细修缮。

老建筑的管线排布比较复杂，为了匹配更多的电器设备，增加空间的使用便利性和后期维修的方便性，还需要进行更缜密的设计，增加安全性。

由于地处繁华市中心，因此对施工过程中的环境污染管控要求也比较高，环保施工、绿色装修以及建筑垃圾的运输等都是整个项目改造中的关键。

即便有诸多困难，设计团队和施工团队还是密切配合，在规定的工期内交出了一份让人满意的答卷。

天津路项目最大限度地尊重现有建筑的原始条件。在没有改变整体结构的前提下加强建筑构造，保留和修复原有墙壁，加入少数新的建筑元素，其理念在于保护历史的经典，注入当代的城市符号及新功能设计。

天津路项目改造中内景
interior of No.66 Tianjin Road project after renovation

The project was originally a three-storey building with an area of 2,546 square meters constructed in 1934, while the fourth floor was constructed in the 1980s, whose construction relationship also needs to be sorted out before careful repair.

The pipeline layout is quite complex. In order to match more electrical equipment, and increase the convenience of space use and later maintenance, more careful design is required to add safety.

The project lies in downtown Shanghai, requiring higher for environmental pollution control in the process of construction. Environmental protection construction, green decoration and the transportation of wastes are crucial during the whole renovation.

Despite numerous difficulties, the design team and the construction team have worked closely together and delivered a satisfactory result within the specified time limit.

The project has maximally respected and reproduced the original building conditions. On the premise of not changing the overall structure, the building structure has been strengthened, the original walls have been retained and repaired, and a few new architectural elements have been added, aiming to inject contemporary urban symbols and new functional design while protecting the historical classics.

天津路项目一层公共空间实景
photo of public space on the ground floor of Tianjin Road project

创新激发城市活力

在室内空间尺度上，营造多层次的错落感。丰富的层次处理，不仅给人以新体验，也是对办公空间差别需求的回应。在公共空间位置保留原有的开敞感，又在办公区域形成了大小不同、多种尺度的独立办公室，以满足不同的企业办公需求。

Stimulation of Urban Vitality with Innovation

A multi-level sense of dislocation has been created in the indoor space, not offering people new experience, but also responding to the different demands for office space. While keeping the original openness in the public space, independent offices of different sizes and scales in the office area have been formed to satisfy the office needs of different enterprises.

天津路项目公共走廊实景照　photo of public corridor of Tianjin Road project

天津路项目公共休息区实景照
photo of public rest area of Tianjin Road project

天津路项目各办公间实景照
photos of office spaces of Tianjin Road project

如何以物质手段体现空间属性是创造个性化空间的出发点。项目在色彩、图案和植物配置方面都做了大胆尝试。改建后室内色彩主调为白色，白墙白顶灰地柔化了空间边界，使空间均质化；而局部搭配浓烈的色彩画面又使空间分异，两极手法的混用，介于二维和三维之间的视觉效果使空间分布别有韵致。

How to embody space attribute by material means is the starting point of creating personalized space. Bold attempts in color, pattern and plant configuration have been made. After renovation, the interior is mainly white: white walls and top surface and gray floor soften the space boundary and homogenize the space, while the local strong colors make the space different, presenting a unique visual effect between two-dimension and three-dimension.

室内空间以白色主基调搭配局部亮色块提升空间韵致
Indoor white main tone with local bright colors to enhance space charm

第三章
CHAPTER THREE

通往独立办公区域的公共通道实景图
real-scene photo of the public channel to the independent office space

地面问题在老建筑的改造修缮中尤其突出，由于长年累月的沉降，建筑楼层地面平衡度差，倾斜严重，同一楼层的高差肉眼可见，有的地方高差竟有十几厘米，只有先解决这个地面平衡度的问题才能铺设其他面材，实现预期效果。

如果采用水泥砂浆找平会大大增加原始钢木结构楼板的承重负担，同时经过局部尝试，也证明自流平达不到理想的找平效果。

经多方考量和测试后，采用了装配式地坪体系。它可以取代细石混凝土找平、自流平等湿作业地面基层，还具有适应性强的优点，能快速铺设，调整后的平整度非常高，即使是地板地砖这种对平整度要求高的材料也可以很服贴。本项目的空间设计以工业风的石塑地板为主，脚感舒适，杜绝安全隐患，同时便于后期局部检修替换。

The ground problem is particularly prominent in the renovation. Due to the long-term sedimentation, the floor is subject to poor ground balance and serious inclination. Visibly, the height difference of the same floor in some places could even reach more than ten centimeters. Therefore, only by solving it can other surface materials be laid to achieve the expected effect.

Using cement mortar for leveling may greatly increase the bearing burden of the original steel and wood floor slab. Besides, through local trials, self-leveling has been proved not able to achieve the ideal leveling effect.

Upon considerations and tests, the prefabricated floor system is adopted. In addition to replacing the fine aggregate concrete for wet operations on the ground such as leveling and self-leveling, it has strong adaptability: the floor can be laid quickly and after adjustment, the flatness is quite high, even on the floor tile, which requires high flatness. The space design of this project is mainly made of SPC floor in an industrial style, which is not only comfortable and can eliminate potential safety hazards, but also convenient for later local maintenance and replacement.

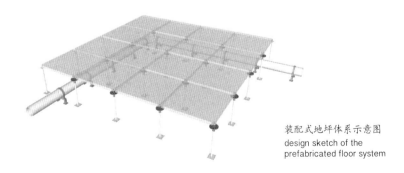

装配式地坪体系示意图
design sketch of the prefabricated floor system

地下室改造后实景图
photo of the renovated basement

地下室从前是银行的金库,如今变成了会员的活动室
basement has changed from a bank vault into a member activity room

墙面采用在原始墙面直铺竹木纤维板的做法，替代内墙涂料或壁纸等。全干法施工大大缩短了施工周期，杜绝了甲醛涂料污染的可能性，保证使用者安全健康。同时竹木纤维板耐刮擦易清洁的特性，给商业空间的日常维护更大的便利性。墙板调平龙骨可以使隐蔽工程更符合实际的使用需要，方便管线排布。办公租户如若在使用过程中对墙面造成局部损毁或难以清除的印迹，因装配式墙面具有局部可拆改的特色，因此后续修缮维护都可轻松处理。

WPC has been directly laid on the original wall to replace the interior wall paint or wallpaper. All-dry construction has greatly shortened the construction period and eliminated formaldehyde paint pollution, ensuring the safety and health of users. In addition, WPC is scratch-resistant and easy to clean, making the daily maintenance of commercial space more convenient. The leveling keel of the wallboard can better align the concealed works with the actual use needs and facilitate the pipeline layout. Even though the office tenant causes local damage or marks that are difficult to remove during the use process, the prefabricated wall allows for local demolition and renovation, so the subsequent repair and maintenance can be easily handled.

墙面处理示意效果图
design sketch of wall processing

墙面实景照
photo of the wall

在企业办公空间中，卫生间的打造也是体现整体品质的关键。本项目采用装配式卫生间系统，可以快速匹配任何尺寸规格的空间，特色防水工艺杜绝渗漏隐患，考究的材料应用可以使水系通达的空间保持干爽且易清洁。视觉上采用白色、绿色和原木色搭配，给人既清爽整洁又时尚的感觉。

In the enterprise office space, the toilet construction is also a highlight that reflects the overall quality. This project adopts the prefabricated toilet system, which can quickly match the space of any size and specification. Its special waterproof process eliminates the potential trouble of leakage. Its exquisite materials keep the space dry and clean. White, green and log colors are matched to deliver people a feeling of fresh, clean and fashion.

整体空间实景效果
real-scene sketch of the interior space

项目交接现场照片　photo of project handover ceremony

设备管线属于隐蔽工程，但其作为"循环系统"直接影响空间的布局，关系到空间使用者的舒适度、检修的方便性及未来的可变性。尤其是老建筑改造成共享办公空间，更多高科技的电器设备入驻其中，用电量需求要扩容，对原本老化混乱的线路系统提出了挑战。

装配式的布线体系与装配式墙顶地面的基层架构相辅相成，能够共同优化升级空间的能量输送系统，局部轻质隔墙内部集中铺设管线是装配式隔墙体系和布线体系完美搭配的效能升级。

Equipment pipelines are one of the concealed works, but as the "circulation system", they directly affect the layout of space, and are related to the comfort of space users, the convenience of maintenance and future variability. In particular, renovating an old building into a shared office space requires more high-tech electrical equipment, so the power demand will increase, posing a challenge to the original aging and chaotic pipeline system.
The prefabricated pipeline laying system and the basic architecture of the prefabricated walls, ceiling, and floor supplement each other, which can jointly optimize and upgrade the energy transmission system of the space. The centralized laying of pipelines in the partial light partition wall is an efficiency upgrade rendered by the perfect match of the prefabricated partition wall system and the pipeline laying system.
The Tianjin Road project has turned out to be a modern shared office community using a set of comprehensive operation and management systems, including real-time operation management system, member management system, sales management system, community management system, procurement management system, financial management system and personnel management system, aiming to create a vibrant and interconnected community. After more than 100 days of renovation, this old building ushered in its new life. At the end of December 2018, the 百空间 Tianjin Road project carefully renovated by Bailian Asset Holding was completed with a new appearance.
All 百空间 members will enjoy the same rights and interests and global community resources as WeWork global members. Boasting unique design and technology highlights, 百空间 will create a humanized work and life style and inspire unlimited creative inspiration.

项目公共空间墙顶布线
photo of ceiling pipelines layout

  天津路项目现已变身成为现代共享办公社区，运用全方位营运管理系统，包含楼宇实时运行状况管理系统、会员管理系统、销售管理系统、社区管理系统、采购管理系统、财务管理系统以及人员管理系统，旨在打造活力四射互相联通的社区。

新风、照明等智能化设施为这个空间创造舒适和便利
intelligent facilities such as fresh air and lighting create comfort and convenience for office space

办公间内配置智能化科技电器
high-tech electrical appliance in the office room

  经历100多个日日夜夜的装修改造，这栋老建筑迎来它的新生，2018年12月底，由百联资控倾心打造的天津路百空间项目以全新的姿态精彩绽放。

  所有百空间会员都将和WeWork全球会员同样享受所有权益与开放强大的全球社区资源链接。百空间凭借独特的设计和科技亮点，打造人性化的工作和生活方式，激发无限创造灵感。

唤醒
RENAISSANCE

项目手记

天津路 百空间 是百联资控首个旧建筑改造项目，整个设计手法简练，充分尊重原有空间结构，取得很好的效果。无论是内部体验，还是城市街景，使用者和公众都给予很高的评价。在可持续发展观念越来越深入人心的今天，珍惜资源成为共识。告别大拆大建的时代，小型的旧建筑改造项目将会越来越多，本项目的诸多创新装修科技运用经验也值得借鉴。

Project Notes

The 百空间 on Tianjin Road is the first old building renovation project of Bailian Asset Holding. Designed simply, on the premise of fully respecting the original space structure, the project has yielded good results. Both users and the public speak highly of the internal experience and the urban streetscape. Today, the concept of sustainable development has been more deeply rooted in the people's mind, making the cherish of resources a consensus. The era of large-scale demolition and reconstruction has passed, there would be more small-scale old building renovation projects, and many innovative decoration techniques applied in this project is also worth learning.

项目现状实景
photo of current project

第四章

经典重启：凯恩宾馆（香港路117号）
—— 新的空间维度，解构建筑生命

Chapter Four

Recurrence of Classics: Kaien Hotel (No. 117 Xianggang Road)

– *Deconstruct the Life of Building at a New Spatial Dimension*

　　在上海的众多中心城区小马路中,香港路是极为低调的,可以说是不大为人熟知。这是一条东西走向的街道,全长仅310米,宽度最窄处9.1米,最宽处也仅有10米。窄窄的马路似乎一伸手就可以够到对面。因为这条路至今不通公交车,便没有了繁华的喧闹。

Xianggang Road is quite low-key and even little-known among many small roads in downtown Shanghai. It is an east-west street, only 310 meters long and 9.1-10 meters wide. On this narrow road, the opposite side seems to be within reach. Without any bus passing through it even today, Xianggang Road is far away from the hustle and bustle.

香港路位于黄浦区东北部，东虎丘路，西至江西中路。19世纪50年代修筑了虎丘路至四川中路段，名诺门路(GnaomenRoad)，清同治元年(1862年)其更名为香港路。约在清光绪二年(1876年)又向东、西两端延筑。香港路地处外滩历史文化风貌保护区内，坐落在道路两旁的诸多优秀历史建筑，就像是海岸线边的一排排连体别墅。

香港路117号紧邻外滩源都市功能区，在黄浦区外滩地区的城市更新及保护性开发背景下，以提升区域产业能级为核心，配套服务于外滩金融集聚带建设大局，积极引进符合外滩金融集聚带功能定位的金融、文化项目或高端服务业总部型企业。

Located in the northeast of Huangpu District, Xianggang Road stretches from Huqiu Road in the east and Middle Jiangxi Road in the west. In the 1850s, the section from Huqiu Road to Middle Sichuan Road was built and named Gnaomen Road. In the first year of Tongzhi in the Qing Dynasty (1862), it was renamed Xianggang Road. In the second year of Guangxu in the Qing Dynasty (1876), it was extended to the east and the west. Xianggang Road is in the Bund Zone with Historical and Cultural Features. Many outstanding historical buildings on both sides of Xianggang Road are like townhouses along the coastline.

No. 117 Xianggang Road is adjacent to the Bund Origin Urban Functional Area. In the context of urban renewal and protective development in the Bund of Huangpu District, headquarters-based enterprises in such industries as finance, culture, and high-end services that meet up to the positioning of this project and the functional positioning of the Bund Financial Agglomeration Belt will be actively introduced to improve the regional industry energy level and support the overall construction of the service area.

香港路路牌
guideboard of Xianggang Road

凯恩宾馆外立面
appearance of Kaien Hotel

香港路街景照片，右侧大楼为香港路59号上海银行公会旧址，尽头是虎丘路上的真光大楼
photo of the past streetscape on Xianggang Road, with former Shanghai Banking Association (No. 59) on the right and China Baptist Publication Building on Huqiu Road at the end of Xianggang Road

采撷不尽的旧日繁花

上海人都知道香港路老房子多，但不太为人所知的是，那些老房子历经岁月沉淀下来的故事更能打动人。

1949年以前，香港路东段排列整齐的大楼内，多为公司写字间和教会办事机构，有中华圣经会、中华基督教青年会等。道路西段多为仓库堆栈，有瑞丰泰堆栈、英商茂泰洋行堆栈和亚洲仓库等。

在香港路上漫步的人也许不会想到100多年前，中国最早的电影公司诞生在这条路上。1909年，亚细亚影戏公司就在香港路1号创建落成，创始人是美籍俄裔商人本杰明·布拉斯基(Benjamin Brasky)。

Countless Old Stories

As known to all Shanghai citizens, there are many old houses on Xianggang Road, but little-known to them, the stories of those old houses are more moving.

Before 1949, most of the buildings in the east section of Xianggang Road were offices of companies and churches such as the Chinese Bible Church and the Young Men's Christian Association. In the west section of the road, there were mainly warehouses, such as Ruifengtai Warehouse, Marden & Co Warehouse, and Asia Warehouse.

People walking on the Xianggang Road may not think that China's first film company was born here more than 100 years ago. In 1909, American Russian businessman Benjamin Brasky founded the Asia Film Company at No. 1 Xianggang Road.

在香港路江西中路路口曾经坐落着上海第一座自来水塔。1883年，英商在上海开办了上海第一家自来水厂，将自来水送往苏州河南岸的租界地区。为了保证日夜都能供水，从杨树浦排水管经江西路桥过河，又特地在桥南的香港路和江西路转角处兴建了一座大型的自来水水塔。通过水塔将经过处理后的自来水输送给每家每户。水厂在夜间不需要工作，凭水塔内的存水供用户使用。在发生火灾时，水塔还能提供大量的消防用水，这在当时的市政建设上具有重大意义。

这座自来水水塔在建成后整整使用了73年。随着上海用水量日益增长，高楼增多，水塔渐渐失去了其功能，于是就被闲置了起来，1954年被拆去。现在香港路已不复见其雄伟的身姿，但它在上海历史上所起的作用却是不可磨灭的。

位于当时江西路香港路的自来水塔历史照片
historical photo of the Tap Water Tower at the
intersection of Jiangxi Road and Xianggang Road

原上海银行同业公会大楼
former Shanghai Banking Association building

The first tap water tower in Shanghai used to be placed at the intersection of Xianggang Road and Middle Jiangxi Road. In 1883, a British businessman founded the first waterworks in Shanghai to send tap water to the concession area on the south bank of Suzhou Creek. In order to ensure water supply day and night, a drain pipe was paved from Yangtszepoo to across the river through Jiangxi Road Bridge, and a large tap water tower was specially built at the intersection of Xianggang Road and Middle Jiangxi Road in the south of the bridge. The treated tap water was supplied to households via the water tower. The waterworks did not need to work at night as the water stored in the water tower was enough. In case of fire, the water tower can also provide a lot of water, which was of great significance in the municipal construction at that time.

After completion, this tap water tower was used for 73 years. With the increase in water consumption and high-rises in Shanghai, the water tower gradually lost its function and was idle; in 1954, it was demolished. Currently, Xianggang Road is no longer magnificent, but its role in the history of Shanghai is indelible.

　　1949年之后，道路沿线仍以仓库、堆栈及商业批发部为多，主要有上海外滩进出口公司仓库、市服装鞋帽公司批发部仓库、市日用杂品公司批发部等。上海市工商界爱国建设公司、国际旅行社上海分社的不少机构也设于此。众多的仓库和堆栈聚集在马路两侧，使原本狭窄的路面越加局促。

　　随着上海产业的变化和经济的发展，香港路上的仓库、堆栈一个个地消失了，路上紧闭着的门面一扇一扇地被打开，楼面上也随之高高悬起了招牌。因为有了爱建的毛纺、工商的纺织品批发部，香港路的服装纺织市场有了一丝成名的先兆。不料，随着苏州河对岸的七浦路服装市场的兴起，香港路刚露头角的服装业之花未开便先谢了。

　　香港路大概真是与热闹无缘，但是也正缘于此，印刻在这条路上的城市记忆也更接近原生态，路上隐藏着的上海传奇也在不经意间染上了一丝神秘的色彩。

　　2007年，上海市城市规划管理局颁布了《关于本市风貌保护道路（街巷）规划管理的若干意见》，将64条马路纳入了一类历史风貌道路保护名单，承诺"永不拓宽"，规定街道两侧的建筑风格、尺度都要保持历史原貌。香港路就列属其中，这些道路分布在上海9个历史文化风貌区，就像流淌着的文脉，串起了建筑经典、历史印痕、人文情怀、生活点滴，受到上海城市"最高级别"的保护。

现存的香港路117号外立面实景（2019年）
real-scene facade of the existing No. 117 Xianggang Road (2019)

After 1949, there were still many warehouses, godowns and commercial wholesale departments along the road, including the warehouse of Shanghai Bund Import and Export Company, the warehouse of the Wholesale Department of Shanghai Clothing, Shoes and Hats Company, and the Wholesale Department of Shanghai Daily Necessities Company. There were also many institutions including Shanghai Patriotic Construction Corporation and Shanghai Branch of China International Travel Service Limited. Numerous warehouses and godowns gathered on both sides, making the narrow road more confined.

With the industrial changes and economic development in Shanghai, on Xianggang Road, the warehouses and godowns have disappeared, the closed doors have been opened, and the signs have been hung on the facades of buildings. Thanks to the Textile Wholesale Department of Shanghai Patriotic Construction Corporation, the clothing and textile market on Xianggang Road showed signs of boom. Unexpectedly, with the rise of the clothing market on Qipu Road by the Suzhou Creek, the clothing industry on Xianggang Road was nipped in the bud.

Probably born to be far away from the bustle, Xianggang Road has carried urban memories closer to the original ecology, and even the legend of Shanghai hidden on it is inadvertently stained with some mysterious color.

In 2007, Shanghai Municipal Planning, Land and Resources Administration issued the *Several Opinions on the Administration of Planning Work for Protective Roads (Alleys/Lanes) with Historical Features in Shanghai*, incorporating 64 roads in the list of protective roads with historical features of grade I which will never be broadened, and the building styles and scales on both sides of them should remain unchanged. Xianggang Road is one of them. These roads are distributed in nine zones with historical cultural features in Shanghai. Like flowing context, they are linked with architectural classics, historical traces, human feelings and life, and are protected at the "highest level" by Shanghai.

邂逅昔日的古典雅致

香港路117号始建于1920年，由德利洋行（Percy Tilley, Graham & Painter Ltd.）的英国建筑师P.Tilley设计。主体为钢筋混凝土框架结构，外观采用古典主义风格，内部则是当时非常先进的多层车库设计，采用升降机提升来实现楼上停车。

项目占地面积为2138平方米，建筑面积为10005.5平方米，古朴庄重颇具欧美风格，是上海市第五批优秀历史建筑，保护类别为三类。

根据本项目查找到历史图纸，该楼原设计为香港路的公共车库，1921年建设完工并开始运营，1941年日本占领时期为女生杂志出版社，1949年后由政府接管后移交上海鞋帽批发部，作为仓库使用。在产证上，它被标记为上海市鞋帽服装公司仓库，隔壁是上海市糖业烟酒公司，这也从侧面印证了香港路在一段时期内曾以服装纺织市场而小有口碑。1994年，楼面部分出租给凯恩宾馆使用，部分为办公。因其坐落于外滩旅游景点西侧的绝好地段，距南京路步行街和乍浦路美食街又很近，因此是早年游客来上海观光旅游下榻的一处好选择。

Encounter with the Classical Elegance in the Past

Built in 1920, No. 117 Xianggang Road was designed by P. Tilley, a British architect of Percy Tilley, Graham & Painter Ltd. The main body is reinforced concrete frame structure. The exterior features classical style, while the interior adopts advanced multi-storey garage design at that time. Lift is applied to realize upstairs parking.

The project covers a floor area of 2,138 square meters and a building area of 10,005.5 square meters. It is simple and solemn in a Western style. It has been listed in the fifth batch of outstanding historical buildings in Shanghai, under protection of Grade III.

According to the historical drawings of this project, the building was originally designed as a public garage on Xianggang Road. It was completed and put into operation in 1921, and was used as a women's magazine press during the Japanese occupation in 1941. After being taken over by the government in 1949, the building became the warehouse of Shanghai Shoes and Hats Wholesale Department. On the house ownership certificate, the building was registered as the warehouse of Shanghai Shoes, Hats and Clothing Company, next to Shanghai Sugar Cigarette & Wine Co., Ltd., also proving that Xianggang Road was once famous in the clothing and textile market in a period of time. In 1994, the building was rented to Kaien Hotel and part of which was used as offices. Located in the west of the Bund tourist attractions and quite close to the Pedestrian Street on Nanjing Road and the Food Street on Zhapu Road, Xianggang Road was a good choice for tourists for sightseeing in the past.

1947年历史行号图
historical commercial map in 1947

香港路117号历史沿革　History of No. 117 Xianggang Road

香港路117号始建于1920年原为Central Gotage Xianggang Road。
No. 117 Xianggang Road was built in 1920 and named Central Gotage Xianggang Road

上海鞋帽批发部在三层屋顶加建一层作为加工生产车间使用西南侧空地有局部增建
Shanghai Shoes and Hats Wholesale Department built an additional floor above the roof of the third floor as a processing and production workshop, with some additional buildings in the southwest open space

1994年，部分出租给凯恩宾馆使用凯恩宾馆对建筑进行局部改造其他部分为办公
In 1994, part of the building was rented to Kaien Hotel and renovated partially for office use

1921年建设完工并开始运营，41年日本占领时期为女生杂志出版社
It was completed and put into operation in 1921, and became a women's magazine press during the Japanese occupation in 1941

20世纪70年代初加建1层
Another floor was built in the 1970s

1994年凯恩宾馆局部改造
In 1994, Kaien Hotel was renovated partially

未来 Future

1920年设计做较大修改
In 1920, the design was greatly modified

1949年由上海政府接受
In 1949, it was taken over by the government

1992年对原建筑加层扩建
In 1992, the original building was expanded

香港路117号始设计初始时间为1919年。英国建设师P.Tilley设计
No. 117 Xianggang Road was designed by P. Tilley initially in 1919

1949年后由上海政府接受并由上海鞋帽批发部使用
After being taken over by the government in 1949, it was used by Shanghai Shoes and Hats Wholesale Department

1992年当时使用单位上海鞋帽批发部委托上海市房产局勘察建筑设计所进行加层改造设计原四层临时建筑整体改造并增加二层加建部分作为办公及招待所使用
In 1992, Shanghai Shoes and Hats Wholesale Department entrusted the Survey and Architectural Design Institute of Shanghai Real Estate Bureau to carry out storey-adding renovation design; the original four-storey temporary building was transformed as a whole and two storeys were added for office and guesthouse use

2019年对香港路117号进行全面修缮
In 2019, No. 117 Xianggang Road will be renovated in an all-round way

1936年拍摄的第一张航拍照片
first aerial photo taken in 1936

1945年拍摄的航拍照片
香港路117号屋顶山花造型
aerial photo of pediment on the roof of No. 117 Xianggang Road taken in 1945

1945年香港路117号
车库车辆提升塔楼
photo of garage vehicle lift tower of No. 117 Xianggang Road taken in 1945

项目外立面是欧洲古典建筑风格,米白色的大楼外观简约明快,线条处理以横线条为主。沿香港路的这一立面装饰比较讲究,底层及屋顶有较深的束腰线,二至三层贯通装饰立柱,局部窗台之间有山花装饰,优雅别致。

香港路117号凯恩宾馆大楼外观
photo of the appearance of Kaien Hotel at No. 117 Xianggang Road

大楼采用三段式立面设计,一层为一个单元,二、三层为一个单元,顶楼为一个单元。东南面及西北面原来分别都设有室外疏散楼梯,现在东南室外楼梯已损毁。主要入口在北立面中间。外墙均为水刷石,有坚如磐石之感觉。门窗采用方形窗,根据车库使用功能,部分窗为装饰窗,不装玻璃,保证车库的通风效果。屋顶用硕大的装饰山墙增加建筑的变化和层次感。

The off-white facade is European classical style, simple and bright, mainly in horizontal lines. The facade along Xianggang Road is decorated exquisitely. There are deep waistlines on both the ground floor and roof, and decorative columns running through the second and third floors. Besides, some windowsills are decorated with elegant and unique pediments.
The facade is divided into three sections: the ground floor, the second and third floors, and the attic. There were outdoor evacuation stairways in the southeast and northwest, but the former of which has been damaged. The main entrance is in the middle of the north facade. The exterior wall uses washed granolithic plaster, as solid as stone. Door windows are square, but according to the use function of the garage, some of them are decorative windows without glazing to ensure the ventilation effect of the garage. The roof is equipped with a huge decorative gable to increase the change and layering of the building.

北立面二至三层两端券窗样式

window pattern on the second and third floors on the north facade

窗台特色花饰

featured floriation on the windowsill

唤醒
RENAISSANCE

新中存旧的跨时空交谈

对这座建筑进行改造，是一个重新认识和发掘的过程，也是一个跨时空重新创造的过程。

从历史图纸得知，本项目在近百年的历史洪流里经历了多次加建，由最初的三层加建至六层，因此四层至六层与整体风格略显不协调。大楼北、西立面为外部重点保护部位，其他原有特色装饰等为内部重点保护部位。

对大楼外立面的改造主要集中在历史风貌的修缮恢复，外墙恢复水刷石表面，加建楼层部分则涂以灰色涂料，并按历史图纸仿制铝仿钢窗，对外墙装饰性的柱帽、窗台、山花进行仔细修缮，维持简洁庄重却又不失细节韵味的风格。另外，首层朝向香港路一侧以大面幅的玻璃窗打造明亮通透的大堂门厅，以提升办公空间的形象和品质感。

▼ 香港路117号项目修缮改造外立面效果图
appearance design sketch of No. 117 Xianggang Road after renovation

Talk between the Old and the New across Time

The renovation of this building is a process of re-recognition and excavation. It is also a process of re-creation across time and space.

It can be seen from the historical drawings that in the past one hundred years, the project has experienced many additional constructions from three floors to six floors, making the style of fourth to sixth floors slightly inconsistent with that of the overall building. The north and west facades are key exterior protected parts; other original characteristic decorations are key interior protected parts.

The renovation of the facade focuses on the restoration of historical features. Washed granolithic plaster will be applied to the exterior wall. The additional floors will be painted with gray coating, and faux aluminum steel windows will be made according to the historical drawings. The decorative column caps, windowsills and pediments on the exterior wall will be carefully repaired to maintain the simple and solemn style without losing the charm of details. In addition, on the ground floor, a bright and transparent lobby hall will be built with large glass windows to escalate the image and quality of the office space.

柱帽修缮效果
repair sketch of the column cap

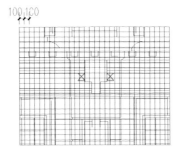
柱帽修缮图纸
repair drawing of the column cap

窗台修缮效果
repair sketch of the windowsill

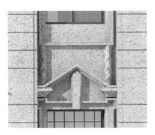
山花修缮效果
repair sketch of the pediment

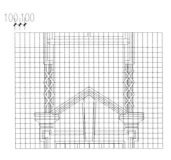
山花修缮图纸
repair drawing of the pediment

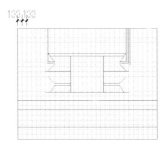
窗台修缮图纸
repair drawing of the windowsill

室内装修方面通过尽可能少的操作，来传达老建筑本身带给场所的静谧感和一种简约的美学。

底层的门厅、二层至六层的电梯厅及公共走道、卫生间都以冷硬利落的基调装修。水磨石地坪、耐磨水泥地坪、清水水泥墙面搭配金属、玻璃、镜面，简洁的边界处理更突出了空间感。

底层门厅采用石膏板吊顶；卫生间做防水石膏板吊顶，利用原设备留孔，对无梁楼板没有破坏；二层以上公共区域平顶检修，涂以灰色涂料；办公区装修则留由租户在未来根据各自需求自行装修。

室内所有公共区域的墙面打磨清理原有涂料，铲除空鼓及损坏部分至基层，用1:1:6混合砂浆修补墙面，涂刷涂料。卫生间新做砖墙。

大堂改造效果图
renovation design sketch of the lobby

Least operations on indoor decorations will be made to convey the sense of tranquility and simple aesthetics brought by the old building itself.

The lobby on the ground floor as well as the elevator halls, public walkways and toilets from the second to sixth floors will all be decorated in a cold and neat tone. The terrazzo floor, wear-resistant cement floor, and fair cement wall will match with metal, glass, and mirror surfaces, whose simple border processing highlights the sense of space.

The lobby on the ground floor will use gypsum board ceiling; the toilets will use waterproof gypsum board ceiling, and holes on the original equipment for avoidance of damage to the girderless floor. The flattops of public areas on and above the second floor will be overhauled and painted with gray coating. The office area will remain original and be decorated by the tenants according to their own needs in the future.

The walls of all indoor public areas will be repainted after polishing and cleaning the original paint, removing the hollowing and damaged parts to the wall substrate, and repairing with 1:1:6 mixed mortar. The bathrooms will be decorated with new brick walls.

本次装修不影响室内原有的无梁楼板柱及柱帽，因此二层以上公共部位柱体仅做清洗处理，露出清水水泥面层。这种"袒露"，就像一个记忆的存储器，将过去的一面真实地展现在未来新用户面前，凝固老的，注入新的，新中存旧，让人可以回味岁月留下的痕迹。

除对大楼本身修缮改造外，本次工程还对总体环境进行整治。在地块内的东侧夹弄，拆除玻璃钢棚屋、废弃管线等物，保留修缮室外楼梯，继续作为垂直疏散通道使用。东侧及南侧通道内现有大量居民搭建，视现场条件拆除。相信这一次改造，将作为一个示范，为整条香港路的更新发展增添一抹亮色。

This decoration will not affect the original girderless floor columns and column caps, so the columns in public parts above the second floor will only be cleaned to expose the surface of fair cement, just like a storage of memories, truly showing the past in front of new users in the future, and allowing them to recall the traces left by the time through a mix of the old and the new.

This project will also renovate the overall environment. In the lane in the east of the block, the FRP sheds and waste pipelines will be demolished, the outdoor stairs will be repaired to continue to be used as the vertical evacuation channel. There are many residential buildings in the east and south sides of the hallway to be demolished according to the on-site conditions. This renovation will serve as a demonstration to add brilliance to the renewal and development of the Xianggang Road.

公共卫生间改造效果图
renovation design sketch of the public toilet

二层走道改造效果图
renovation design sketch of the hallway on the second floor

电梯厅改造效果图
renovation design sketch of the elevator hall

项目手记

"外滩历史文化风貌区"是上海城市的珍贵遗产。香港路上一幢幢经典的建筑,诉说着历史的故事,又上映着当代城市的经典剧目。

通过对香港路117号项目的保护修缮,必将从城市整体环境、城市建筑形态到大楼内在品质上,对保护、提高外滩历史文化风貌区的整体品质起到积极和有效的促进作用。此项目的保护性修缮按照上海外滩历史建筑"重现风貌,重塑功能"的政府保护开发理念,在严格遵守和切实满足建筑保护要求的前提下,结合运营需求、有机更新建筑功能,做到了更有效地促进历史建筑保护和城市功能开发的协调性和统一性。

Project Notes

The Bund Zone with Historical and Cultural Features is a precious heritage of Shanghai. The classical buildings on Xianggang Road are telling the story of history and showing the classical play of a modern city.

The protective repair of No. 117 Xianggang Road project will play a positive and effective role in promoting the overall quality of the Bund Zone with Historical and Cultural Features from the overall urban environment, urban building forms, and the internal building quality. In line with the government's protective development concept of "reappearing features and reshaping functions" of historical buildings in the Bund, on the premise of strictly obeying and practically satisfying the protective repair requirements, coupled with the operation needs and in order to organically update its functions, the protective repair of this project has more effectively promoted the coordination and unity between the protection of historical buildings and the development of urban functions.

*注:本章节部分历史资料原引自朱惜珍著《永不拓宽的上海马路(一)》,东方出版中心2016年7月出版。

*Note: Part of the historical materials in this chapter were originally quoted from *The Roads Never Broadened in Shanghai (I)* by Zhu Xizhen, Orient Publishing Center, July 2016.

第四章
CHAPTER FOUR

香港路117号建筑细节（新旧融合）
building details of No. 117 Xianggang Road (mix of old and new elements)

第五章

文化融合：贵州路263号（原陈英士纪念堂）
—— 从老建筑的新生看历史街区的更新

Chapter Five

Cultural Integration : No. 263 Guizhou Road (Former Chen Yingshi Memorial Hall)

– Renewal of Historical Blocks from the Perspective of the Rebirth of Old Buildings

北京东路是沉寂的，因为这条路上鲜见商场，与隔路的南京东路相比，这条路犹如山林隐士，但它曾经的显赫身世并没有湮没在上海滩的滚滚红尘之中。

但时至今日，北京东路面临着城市发展中的诸多问题：缺乏门户形象，业态老化，几经改造的无序立面及稍显混乱的路面交通，等等。针对现状，近年来黄浦区以城市更新为抓手，围绕"苏河南岸、海派硅巷"主题定位，通过实施"重塑功能、重现风貌、重构产业"的"三重战略"，注入新兴产业、促进功能复合提升。

通过社区改造升级、沿线重要历史建筑业态重置更新等方式，北京东路地区将被打造成中心城区的创新高地、文化高地、保护高地，成为24小时活力街区，赋予历史街区新的生命力。

具有代表性的新生老建筑，是位于黄浦区贵州路北京东路路口的贵州路263号项目。以城市更新建设为支撑，推进创新、创业基地建设，打造成为创意办公和零售商业相结合的复合业态。贵州路263号项目作为 BA/work 系列产品，将一批充满活力的人引入这片街区内工作与生活。随着更新的深入，它将有效提升周边街区的品质及商业配套的完善，助力营造可漫步、可阅读、有温度的街区。

The East Beijing Road is quiet as there are few shopping malls. Compared with the Middle Nanjing Road only a road away, it is like a hermit in the mountain, but its distinguished origin has not lost in the past of the Bund.
But up to now, the East Beijing Road is facing many development problems: lack of portal image, aging business forms, disordered facades after several renovations, slightly chaotic road traffic, etc. Therefore, in recent years, with focus on urban renewal, Huangpu District has been introducing emerging industries and promoting the composite ascension of functions by implementing the strategy of "reshaping functions, reappearing features, and reconstructing industries" around the thematic positioning of building "a silicon alley on the south bank of Suzhou Creek".
By such means as renovating and upgrading the communities and replacing the business forms of old buildings, the East Beijing Road will be built into an innovation highland, a cultural highland and a protection highland in downtown Shanghai, and becoming a dynamic block injected with new vitality.
A representative reborn old building is No. 263 Guizhou Road at the intersection of Guizhou Road and East Beijing Road in Huangpu District. The construction of an innovation and entrepreneurship base has been promoted with the support of the construction of urban renewal to create a composite business form combining creative office and retail commerce. As one of BA/work series products, No. 263 Guizhou Road will introduce a group of energetic people to work and live in this block. The deepening urban renewal will effectively improve the quality of surrounding blocks and perfect commercial supporting facilities to help build a walkable, readable and warm block.

走进百年历史的底蕴

　　北京东路是黄浦区北部的一条东西走向的道路，东起中山东一路（外滩），西至西藏中路，全长1683米，宽21~27米。该路始建于1849年，最初名为领事馆路，因东端设有英国领事馆而得名，俗称"后大马路"；1854年向西筑至浙江北路；1865年以清廷京城命名为北京路；1876年又延伸到西藏中路；1945年更名为北京东路。

19世纪末拍摄的北京路外滩西望
historical photo of Beijing Road in a west view from the Bund at the end of the 19th century

1913年商务印书馆实测上海城厢租界地图中绘制的北京路及周边区域
Beijing Road and its surrounding areas in the survey map of Shanghai County and Concession Areas by the Commercial Press in 1913

Tap into the Essence of the Centurial History

The East Beijing Road is an east-west road in the north of Huangpu District. It is 1,683 meters long and 21-27 meters wide, stretching from First East Zhongshan Road (the Bund) in the east and Middle Xizang Road in the west. Built in 1849, it was initially named Consulate Road after the British Consulate in the east, and commonly known as the Back-Middle Road. In 1854, the East Beijing Road was built westward to North Zhejiang Road and named Beijing Road in 1865 after the capital of the Qing Dynasty. In 1876, it was extended to Middle Xizang Road and in 1945, it was again renamed East Beijing Road.

第五章
CHAPTER FIVE

北京东路贵州路口实景 (2019年)
photo of the intersection of East Beijing Road and Guizhou Road (2019)

北京东路是一条有历史底蕴的道路，在东段鳞次栉比地坐落着一幢幢气度不凡的银行大楼，半部中国银行史留存在这条道路上。格林邮船公司大楼及怡和洋行大楼、上海友谊商店、益丰洋行大楼、金城大戏院（现黄浦剧场）……一个个耳熟能详的名字对上海市民来说都是一处处记忆深刻的地方。

The East Beijing Road boasts historical profundity. In the east section, there are many bank buildings with extraordinary bearing and half of the history of banks in China remains on it. Glen Line Building, Jardine Matheson & Co. Building, Shanghai Friendship Store, Abraham Co. Building, the Lyric Theatre (now Huangpu Theater) … These familiar names have left deep memories for Shanghai citizens.

1987年拍摄的北京东路西望 现为益丰外滩源
East Beijing Road from a west view in 1987, now it is Yifeng Galleria

金城大戏院开张不久后的影像
photo of the Lyric Theatre after its opening

20世纪早期拍摄的外滩怡和洋行老楼
old Jardine Matheson & Co in the Bund in early 20th century

尤其是黄浦剧场，曾在上海黄浦人民文化生活里有着重要地位。黄浦剧场的前身是金城大戏院，始建于1933年，1934年2月1日正式开业。早期曾有《渔光曲》《风云儿女》等大量国产优秀影片在此首映，被誉为"国片之宫"。1935年，国歌《义勇军进行曲》首次在此唱响并传播。1957年，周恩来总理提议更名为"黄浦剧场"并亲笔题字。

黄浦剧场日景 daytime scenery of Huangpu Theater

Especially, Huangpu Theater once played an important role in the cultural life of people in Huangpu District. Its predecessor is the Lyric Theatre founded in 1933 and officially opened on February 1, 1934. In the early stage, a large number of excellent domestic films such as the *Song of the Fishermen Lyrics* and the *Children of Trouble Time* were premiered here, making the Lyric Theatre win the title of "The Palace of National Films". In 1935, the national anthem *March of the Volunteers* was first sung and spread here. In 1957, Premier Zhou Enlai proposed to rename it Huangpu Theater and wrote his inscription.

经历一百多年的历史，北京东路周边已形成了内容丰富的历史街区，老建筑背后都有着深厚的底蕴。

With a history of more than one hundred years, a rich historical block has been formed around the East Beijing Road, on which the old buildings bear historical profundity.

倾听旧时光的衷肠

贵州路623号位于北京东路和贵州路路口,距离南京东路商圈仅500米,所处地段可谓寸土寸金。

项目始建于1930年,建筑面积为3187.85平方米,为四层框架结构建筑。南侧分布五金机电类批发市场,东侧毗邻黄浦剧院,北侧和西侧环绕着许多老旧住宅小区。改造前,一层由全家便利店及五金类零售商铺构成,二层至四层作为办公功能使用。

该项目原为陈英士纪念堂,初始为三层建筑,经过历年改建扩建,现存为四层建筑,仍尚存最初的历史风貌。比对历史照片,沿街外立面的二层至四层基本保留了原窗洞布局及比例和原立面线脚的装饰。

Listen to the Touching Story in the Past

No. 623 Guizhou Road is located at the intersection of East Beijing Road and Guizhou Road, only 500 meters away from the business district of East Nanjing Road. It lies in an area with rent of top high.

Built in 1930, the project is a four-storey frame structure building, with a building area of 3,187.85 square meters. It has a hardware and electrical wholesale market in the south, neighbors Huangpu Theater in the east, and many old residences in the north and west. Before the renovation, the ground floor consisted of FamilyMart and hardware retail shops, while the second to fourth floors were used for office.

The project was the former Chen Yingshi Memorial Hall, initially a three-storey building. After years of reconstruction and expansion, it has become a four-storey building with its original historical features. According to the historical photos, the layout and proportion of the original window openings from the second to fourth floors on the facade along the road, and the decorations of the architraves on the original facades have been basically preserved.

现状立面图
Current facade

现状建筑细节 details of the current building

第五章
CHAPTER FIVE

贵州路623号大楼外观
appearance of the building at No. 623 Guizhou Road

唤醒
RENAISSANCE

室内部分，在现存主入口进入的大堂区域，双分对折楼梯以米白色大理石及米黄色水磨石为主体饰面材料，配之以深灰色水磨石线脚装饰物，用材做工考究，且保存完整，置身其中仿佛穿越了时空过往，倾听着旧时光的衷肠。

邻近东北角一侧的楼梯间也保留着历史原貌，木质扶手、铁艺护栏、米黄色水磨石梯段，每块材料都充满了那个年代的真实感。

In the lobby which can be entered via the extant main entrance, the double split half-fold stairwell uses off-white marble and beige terrazzo for main facing, with dark gray exquisite and well-preserved terrazzo architrave ornaments. It seems that you are listening to a past touching story across time.
The stairwell near the northeast corner also retains its original appearance. Wooden handrails, iron guardrails, beige terrazzo stairs are all full of the sense of reality of that era.

贵州路项目现状实景照片　　　　　　　　　　　装饰主义立面现状
photo of Guizhou Road project　　　　　　　　current appearance featuring Art Deco style

贵州路项目现状实景照片
photo of Guizhou Road project

石面楼梯现状
current stone-facing stairs

楼梯雕花护栏现状
current carving guardrails of stairs

拼花大理石
parquet marble

1878年1月17日,陈其美(字英士,号无为)出生在浙江湖州吴兴的一个商人家庭。读过七年私塾,在当铺当过十二年学徒,后到上海做了两年同康泰丝栈的佐理会计。

1906年,他在弟弟陈其采的资助下东渡日本留学,同年冬天加入中国同盟会,从此走上了近代民主革命之路。在日本,他结识了蒋介石,并介绍蒋加入了中国同盟会,引导其参加辛亥革命,可以说是蒋介石的"领路人"。

1908年回国后,他在上海的革命活动中崭露头角,吸引了孙中山的注意。他结交青帮以掩护革命活动,吸收大批江浙资本家加入同盟会,结识商界名人、社会名流,推动他们赞助革命活动,使中国同盟会在上海有了比较扎实的社会基础。在辛亥革命初期,陈其美与黄兴同为孙中山的左右股肱,被蔡元培盛赞为"民国第一豪侠"。

袁世凯上台后,陈其美追随孙中山进行二次革命和护国战争,曾策划一系列反袁军事行动。1916年,受袁世凯指使的张宗昌派人,假借签约援助讨袁经费,在日本人山田纯三郎寓所中将陈其美暗杀。

陈其美死后,各地陆续修建了陈英士纪念设施,南京建有英士图书馆,湖州开办了英士学校,杭州设立陈英士专祠,唯独上海没有纪念设施。1929年5月,陈英士三弟陈蔼士偕同湖州同乡会数人致电当时的国民政府,力陈"上海为陈公英士生前开府之所,捐躯殉国之地,而纪念建筑迄今尚付阙如,似不足以慰先烈而示来兹。"他建议政府批准并资助湖州旅沪同乡会"湖社"来建立陈英士纪念堂。这一建议为国民政府接纳,通过中央和地方政府拨款、资助,最终建成了上海英士纪念堂。

老照片中可看到雕刻了"陈公英士纪念堂"字样的门头
front with written characters of "Chen Yingshi Memorial Hall" in an old photo

On January 17, 1878, Chen Qimei (whose courtesy name is Yingshi and art name is Wuwei) was born in a merchant family in Wuxing District, Huzhou City, Zhejiang Province. He learned in a private school for seven years, served as an apprentice in a pawnshop for twelve years, and later worked as an assistant accountant in Tongkangtai Silk Shop in Shanghai for two years.
In 1906, with the financial help of his younger brother Chen Qicai, Chen Qimei went to Japan for studies. In the winter of that year, he joined the Chinese Revolutionary League, embarking on the road of modern democratic revolution. In Japan, Chen got to know Chiang Kai-shek. Chen Qimei introduced Chiang Kai-shek to join the Chinese Revolutionary League, and led him to participate in the 1911 Revolution. He was an early mentor of Chiang Kai-shek.

贵州路项目现状实景照片
photo of Guizhou Road project

After returning to China in 1908, Chen Qimei cut a figure in revolutionary activities in Shanghai and attracted Dr. Sun Yat-sen's attention. He associated with the Green Gang to shield the revolutionary activities, attracting many capitalists in Jiangsu and Zhejiang to join the Chinese Revolutionary League. Chen Qimei made friends with commercial and social celebrities and persuaded them to sponsor the revolutionary activities, ensuring a relatively solid social foundation of the Chinese Revolutionary League in Shanghai. In the early period of the 1911 Revolution, Chen Qimei and Huang Xing were both capable assistants to Sun Yat-sen. Chen Qimei was praised by Cai Yuanpei as the "Hero of the Republic of China".

After Yuan Shikai came to power, Chen Qimei assisted Sun Yat-sen in the second revolution and the national defense war, and planned a series of military actions against Yuan Shikai. In 1916, instructed by Yuan Shikai, Zhang Zongchang assigned someone to assassinate Chen Qimei in the house of Junichiro Yamada under the guise of signing a contract to fund for suppressing Yuan.

After his death, memorial facilities for Chen Yingshi were built in succession all over the country, including Yingshi Library in Nanjing, Yingshi School in Huzhou, and Yingshi Ancestral Hall in Hangzhou, but no memorial facilities were built in Shanghai. In May 1929, Chen Aishi, the third younger brother of Chen Yingshi, together with several members of Huzhou Natives Association, called the national government, urging that "Shanghai was where Chen Yingshi lived and died for, but no memorial has been built yet, which seems to be not enough to show sympathy." He suggested that the government approve and fund the Huzhou Natives Association to build Chen Yingshi Memorial Hall. This proposal was accepted, and Chen Yingshi Memorial Hall was finally built with funding from central and local governments.

打造"海派硅巷"

结合大量海内外类似建筑经验以及项目现存状况,百联资控提出了保留主要结构,立面及室内进行调整和美化的改造策略,从实用性、经济性的原则出发,尽量沿用现有平面布局及现有设备如电梯,机房,水箱等。在此基础上结合地块规划方案粉刷更新立面,修复及清洗过于老旧的门窗,使整个建筑拥有基本良好的空间体验。

Create a "Silicon Alley in Shanghai"

Combined with a large number of similar construction experience at home and abroad and the status quo of the project, Bailian Asset Holding has put forward the renovation strategy of retaining the main structure, and adjusting and beautifying the facades and interior. In line with the principle of practicability and economy, the extant plane layout and equipment such as electric ladder, machine room, and water tank would be retained. On this basis, coupled with the plot planning scheme, the facades would be painted and renewed, and the old door windows would be repaired and cleaned, so that the whole building will have a basically good space experience.

项目改造效果图
renovation design sketch of the project

业态全面调整，打造创意办公联动底层商业模式

作为 BA/work 系列产品，本项目一层在延续沿街面商业集聚氛围的同时，提供服务于办公客群、黄浦剧场观众及周边社区居民消费和生活休闲的社交环境。二层至四层功能置换为创意办公空间，吸引创新的年轻办公客群，为楼下商业带来了消费活力。

立面及室内公区全面升级，带动整体街区品质提升

为表达对历史的记忆与尊重，整个建筑的经典元素得到保留，创造一种时间流动穿梭的氛围，同时空间嵌入新兴的功能，提升街区品质。

在外墙立面部分，设计团队一方面保留项目原立面线脚，立面整体重新粉刷真石漆；另一方面，增加立面玻璃采光面积，增设沿街东侧及南侧办公入口，通过更新整体照明系统来提高陈旧建筑的开放型新形象，增加现代时尚元素。

Comprehensive adjustment of business forms to create the creative office linkage business model on the first ground floor

This project is one of the BA/work series products. While continuing the atmosphere of business cluster along the road, the ground floor will provide a social environment for office customers, audience of Huangpu Theater, and residents of the surrounding communities to consume, live and relax. From the second to the fourth floors, the function will be replaced with creative office space to attract innovative young office customers and bring consumption vitality to the downstairs commerce.

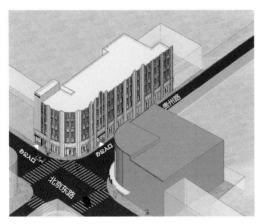

项目改造设计效果图 renovation design sketch of the project

[Comprehensive upgrading of the facades and indoor public areas to drive the quality enhancement of the whole block]

In order to express the memory and respect for history, the classical elements of the building will be preserved to create an atmosphere of time flowing and shuttling, while new functions will be embedded in the space to enhance the quality of the block.

In terms of facades, the design team on the one hand retains the original architraves and repaints with stone-like coating; on the other hand adds the glass lighting area and office entrances on the east and south facades along the street, improves the open new image of the old building by updating the overall lighting system, and incorporates modern fashion elements.

内部空间主要通过电梯系统、机电系统、空调系统的升级来实现物业品质升级。一层商业空间及室内公共区域将重新翻修，在保护建筑自身特色并且给予租户较大自由度的同时，延续BA/work产品线简约一致的风格。

The internal space mainly upgrades the property quality by upgrading the elevator system, electromechanical system and air conditioning system. The commercial space and indoor public area on the ground floor will be renovated. While protecting the building characteristics and giving tenants more freedom, the simple and consistent style of BA/work project lineups will be continued.

外立面设计图 Design sketch of appearance

室内公区装修效果图
decoration design sketch of the indoor public area

第五章
CHAPTER FIVE

项目改造效果图
renovation design sketch of the project

穿越历史 邂逅经典

项目手记

在本项目的改造中，面临着原建筑的历史性如何与周围街区的功能重塑、风貌和产业升级协同发展的命题。人们需要记忆，城市也需要记忆。充满人文故事的老建筑毫无疑问是城市文化、记忆的最佳载体，也是抵抗时间留下审美、技术、精神等财富的容器。但老建筑也要承担被时间裹挟向前做出改变、发展的命题。我们看到了贵州路263号项目在这一命题上的作答。城市更新不仅是商业价值的提升，也是让建筑带着它们的故事继续前行，让人们依旧可以触摸到城市记忆的温度。

Project Notes

In the renovation of this project, there is a problem of how the historic property of this building can develop together with the function remodeling, features, and industrial upgrading of the block it belongs. People need memory, so does a city. Undoubtedly, the old buildings bearing countless human stories are the best carrier of urban culture and memory, as well as a container of treasures such as aesthetics, technology and spirit left in the past. However, the old buildings also have to bear the proposition of making changes and development under the pressure of time. We have seen the answer of No. 263 Guizhou Road Project on this proposition. Urban renewal is not only the promotion of commercial value, but also the continuous progress of buildings with their stories, so that people can still touch the temperature of urban memory.

* 注：本章节部分历史资料参考：庄智娟. 湖社和陈英士纪念堂[J]. 档案春秋, 2009年02期.

Note: Part of historical materials in this chapter refer to: *Hunan Natives Association and Chen Yingshi Memorial Hall* by Zhuang Zhijuan, published in *Memories and Archives*, Vol. 02, 2009.

贵州路项目现状实景照片
photo of Guizhou Road project

第二部分 PART 2

北外滩金融产业办公集群——引领区域产业升级
【新城市的价值再建与产业重构】

North Bund Financial Industry Office Cluster: Lead the Regional Industrial Upgrading
Value Reconstruction and Industry Restructuring of a New City

唤醒
RENAISSANCE

现代意义上的城市更新，已不仅仅局限于改善城市物质环境，还有更广泛的社会与经济复兴重任在身。

当我们现在谈论北外滩时，联想到的是她错落的天际线，以及充满年代感的码头和老建筑。可曾想二十几年前的北外滩，这里还是以石库门老房子为主的片区，楼高28层的远洋宾馆在其中显得鹤立鸡群。而现在，320米高的白玉兰广场摘下"浦西第一高楼"的桂冠，拥有"一滴水""彩虹桥""音乐之门"三大特色景观的北外滩滨江绿地被称为"最美江景步道"，众多大型国有企业和金融机构纷纷入驻。

第二部分 PART 2

In modern sense, urban renewal is not limited to improving the physical environment of a city, but also bears a more extensive task of social and economic rejuvenation.
When we talk about the North Bund, we think of its scattered skyline, old wharfs and historical buildings. But more than 20 years ago, it was dominated by Shikumen old houses, among which the 28-storey Ocean Hotel stood out. Now, Shanghai Magnolia Plaza, which is 320 meters high, has been hailed as the "Highest Building in Puxi". The North Bund Green Land with three feature landscapes is called "the most beautiful riverside footpath". Many large state-owned enterprises and financial institutions have settled in.

北外滩鸟瞰图
aerial view of the North Bund

占地3.66平方公里、浦江风景线长达3.53公里的北外滩，紧邻外滩、与陆家嘴隔江相望，成为上海最炙手可热的黄金板块，被贴上了"第二个陆家嘴"的标签。作为外滩的延伸，这里拥有丰富的文化资源和深厚的历史积淀，坐北朝南，面水朝阳，是上海航运业的发源地。经过二十多年的磨砺发展，航运服务业已成为虹口区的一张名片，从经济层面来看，其支柱产业的地位也非常明显。

如今，关于北外滩的"造势"正从滨江一线逐渐向内移，从"航运聚集区"发展成为"航运与金融服务聚集区"。

"北外滩的美好前景，一定是建基于它丰富灿烂的人文历史和独具优势的自然条件上。希望能够通过各方面的共同努力，从规划入手'尽力做到现代化建筑和丰厚的文化底蕴完美结合'，避免再出现各种遗憾。"百联资控CEO程大利女士这样表示。

未来五年，北外滩南部地区将基本完成旧区改造，其核心区将主要打造成商务文化街区，新增商务商业载体190万平方米。此前，虹口规划部门曾公布，北外滩将打造一条"空中绿街"，东至星港国际中心、西至白玉兰广场，向南直抵滨江绿地，串起多个历史文化街区。

在北外滩片区众多建筑中，百联资控打造的是两座比较典型的老建筑——杨树浦路61号和147号，修缮完成后将充分发挥其现代文化价值和经济价值，催生北外滩对中高商务办公的强烈需求，共同构筑北外滩的绿色商办建筑群。

The North Bund, which covers an area of 3.66 square kilometers and has a scenery line of 3.53 kilometers, is close to the Bund and faces Lujiazui across the river. It has grown into the hottest golden block in Shanghai with the label of "another Lujiazui". As an extension of the Bund, the North Bund boasts abundant cultural resources and historical profundity. Facing south and the sun, it is the birthplace of the shipping industry in Shanghai. After more than two decades of development, the shipping service industry has become a business card of Hongkou District. From the economic perspective, the status of its pillar industry is also obvious.
With momentum moving inward from the riverside, the North Bund has been developing from a cluster of shipping to a cluster of shipping and financial services.

百联资控团队在杨树浦路61号项目现场考察　　　　　　　　　　　　　　　　　　　杨树浦路61号现状实景
on-site survey of No. 61 Yangshupu Road project by Bailian Asset Holding team　　　photo of the current No. 61 Yangshupu Road

"The bright prospect of the North Bund must be based on its rich and splendid human history and unique natural conditions. We hope that through joint efforts, we can try our best to perfectly combine modern building and rich cultural heritage from planning, so as to avoid all kinds of regrets", said Ms. Cheng Dali, CEO of Bailian Asset Holding.

In the next five years, the south part of the North Bund will basically complete the reconstruction of old area, build its core area into a business and cultural block, and add 1.9 million square meters of business and commercial carriers. The planning department of Hongkou announced that the North Bund will be built into a "Green Street in the Sky" stretching to Star Harbor International Centre in the east, Magnolia Plaza in the west, and Green Land in the south, on which there are a few historical and cultural blocks.

Among the numerous buildings in the North Bund, two typical old buildings, No. 61 and No. 147 Yangshupu Road have been renovated by Bailian Asset Holding. After renovation, they will give full play to their modern cultural and economic value, promote the strong demand of the North Bund for medium and high business office, and co-build a green business office building complex in the North Bund.

第六章

梦想起航：永兴仓库（杨树浦路61号）
—— 发挥时代功能的空间再造

Chapter Six
Sail the Dream: Yung Shing Warehouse (No. 61 Yangshupu Road)
– *Function the Space Reconstruction of the Times*

第六章 CHAPTER SIX

说到"远东第一"这四个字，你第一时间会想到什么？是苏州河畔上海邮政博物馆的远东第一厅？是长阳路上的远东第一监狱提篮桥监狱？还是远东第一公寓河滨大楼？亦或是远东第一大饭店浦江饭店？

北外滩的远东第一实在太多了，以至于你很难想到"远东第一库"永兴仓库，这座老建筑就位于杨树浦路61号，至今已经有近90年的历史。建筑风格上是典型的巴洛克风格，经历了岁月的洗礼，依旧给人惊喜。

When it comes to "No.1 in the Far East", what do you think of at the first time? The Shanghai Postal Museum on the bank of Suzhou Creek? The Tilanqiao Prison on Changyang Road? The Riverside Building? Or the Astor House Hotel?

There are so many outstanding buildings in the North Bund that you may hardly think of Yung Shing Warehouse, the No.1 Warehouse in the Far East. Located at No. 61 Yangshupu Road, this old building has a history of nearly 90 years in a typical Baroque style, even stunning today.

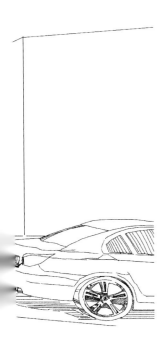

杨树浦路61号建筑外观局部

part of the appearance of No. 61 Yangshupo Road

1937年，这里曾经是"席梦思"公司的所在。1937年"八·一三"淞沪战争爆发后，上海除租界以外全部沦陷，美商席梦思公司处于停滞状态。1942年12月7日，太平洋战争爆发，美国政府向日本宣战，席梦思公司退出上海，杨树浦路61号变成了仓库。

1937年登记的"席梦思公司"在杨树浦路61号
Simmons Company registered at No. 61 Yangshupu Road in 1937

1947年地图标示原来的席梦思公司变成了永兴货栈
former Simmons Company became Yung Shing Godown in the map of 1947

In 1937, Simmons Company was located here. After the Battle of Shanghai broke out on August 13, 1937, Shanghai was occupied except for the concessions, and Simmons Company was in a state of stagnation. On December 7, 1942, the Pacific War broke out. The U.S. government declared war on Japan. Simmons Company withdrew from Shanghai, and No. 61 Yangshupu Road became a warehouse.

1992年前后历史照片，距离远东饭店不远处右下角即为本项目
historical Photo in 1992, with this project in the lower right corner, not far from the Far East Restaurant

杨树浦路61号外立面特写
close-up photo of the appearance of No. 61 Yangshupu Road

唤醒 RENAISSANCE

未来，这栋建筑将承载新的历史使命。第九届北外滩财富与文化论坛上，拟建于永兴仓库的上海金融科技园区正式揭幕。"上海金融科技园区"将以金融行业信息化需求为导向，以前沿高科技应用为支撑，通过构建符合金融科技产业要求的高科技基础设施体系、金融资源投融资对接体系、产业扶持政策体系、专业人才服务体系、国际交流合作体系等集聚国内外优质金融科技企业和相关功能性机构，着力培育金融科技研发、展示、成果转化和资本运作等高端产业链。

拟成为上海金融科技园区的杨树浦路61号
No. 61 Yangshupu Road planned to be built into Shanghai Financial Science and Technology Industrial Park

In the future, this building will carry a new historical mission. At the 9th North Bund Fortune and Culture Forum, the Shanghai Financial Science and Technology Industrial Park to be built at Yung Shing Warehouse was officially inaugurated. Oriented by the informatization needs of the financial industry and supported by cutting-edge high-tech applications, Shanghai Financial Science and Technology Industrial Park will be dedicated to cultivating high-end industrial chains such as R&D, exhibition, achievement transformation, and capital operations of financial science and technology by pooling quality financial science and technology enterprises at home and abroad through the construction of high-tech infrastructure system, financial resources investment and financing docking system, industrial support policy system, professional talent service system, international exchange and cooperation system that meet the requirements of the financial science and technology industry.

杨树浦路61号修缮完成后，建筑面积近37360平方米，将成为BA/work产品线中非常重要的一个项目，届时预计可容纳100～200家优质金融科技企业入驻。根据计划，该园区建成后致力于在2～3年内构建起科技、金融机构有机统一、互动发展，先进技术产业和高效金融资本流动相辅相成的"金融科技生态圈"。园区将成为进一步深化服务业扩大开放的综合试点，争取上海新一轮金融开放政策在园区先行先试，为北外滩财富管理高地不断提升能级，为上海建设具有全球影响力的金融中心和科技创新中心提供强有力支撑。

The renovated No. 61 Yangshupu Road covers a building area of nearly 37,360 square meters, and will become an important project in the BA/work project lineup, which is expected to accommodate 100-200 high-quality financial science and technology enterprises. According to the plan, the park is committed to building into a financial science and technology ecosystem in which science and technology and financial institutions are organically unified and interactively developed, and advanced technology industries and efficient financial capital flows complement each other within two to three years. The park will become a comprehensive pilot to further deepen the opening-up of the service industry, striving to pioneer the new round of financial opening policy in Shanghai, continuously enhance the energy level of the North Bund as a wealth management highland, and provide strong support for Shanghai to build into a financial center and a technological innovation center with global influence.

杨树浦路61号及北外滩实景
photo of No. 61 Yangshupu Road and the North Bund

杨树浦路61号实景
photo of No. 61 Yangshupu Road

浓荫掩映下的杨树浦路61号 ▶
No. 61 Yangshupu Road under the shade of trees

唤醒
RENAISSANCE

曾经的"远东之最"

杨树浦路位于虹口区东南部和杨浦区南部的滨江地带，西起惠民路，东至黎平路，路长5586米，是上海工业的发源地，曾经是"沪东第一路"。

上海开埠后，上海公共租界工部局在黄浦江边修了一条沿江马路，通到"杨树浦"，因此得名"杨树浦路"。这是一条绵延5公里的百年工业之路，沿路北侧为住宅区，有鼎和里、新康里、华忻坊等26条旧式里弄；南侧为工厂区，有毛条厂、毛纺厂、丝织厂、船厂、发电厂、煤气厂、水厂、化工厂、制皂厂等40多家大中型工厂。150年前，工人们每天穿过杨树浦路到工厂去上班，下班又回到马路对面的居民区，"南厂北住"的杨树浦路由于最早拥有完善的市政配套管网，被誉为"沪东第一路"。

1908年，这条路上出现了70辆英国有轨电车，车身是印度红，金黄色嵌线。

Past Fame in the Far East

Located in the riverside region in the southeast of Hongkou District and the south of Yangpu District, Yangshupu Road is 5,586m long, stretching from Huimin Road in the west and to Liping Road in the east. It is the birthplace of the industry in Shanghai and used to be "The No.1 Road in the East of Shanghai".
After the opening of Shanghai, Shanghai Municipal Council built Yangshupu Road named after its stretching to Yangtszepoo by the Huangpu River. It is five kilometers long. The north part is the residential area, including 26 old-style alleys such as Dinghe Alley, Xinkang Alley and Huaxinfang. In the south part, there are more than 40 large and medium-sized factories engaged in wool top, wool spinning, silk, ships, electricity generation, gas, water, chemicals, and soap. 150 years ago, workers crossed Yangshupu Road for work every day, and returned to the residential area opposite the road after work. Hence, Yangshupu Road was hailed as "The No.1 Road in the East of Shanghai" because it was the first to have a complete municipal supporting pipeline network.
In 1908, 70 British tramcars in Indian red with gold inlaid lines ran on Yangshupu Road.

历史老照片：杨树浦路上的有轨电车
historical photo: tramcar on Yangshupu Road

杨树浦路61号项目原为"永兴仓库",是一幢高22.6米六层楼的仓库建筑,占地面积6227平方米,总建筑面积37362平方米。该仓库大楼有如此庞大的建筑体量,宛如一座坚固的城堡,伫立在杨树浦路、惠民路及汇山码头之间的航运要冲地区,区位十分优越。

"永兴仓库"是上海仓储业中最有名的仓库之一。1929年,康益洋行先在倍开尔路(今惠民路36号)上建造一幢四层钢筋混凝土结构的仓库,从事货栈仓储业务。该仓库位于虹口、杨浦交界,是上海早期工业、贸易、航运业最集中的地段,以致常出现仓储需求量大于可容仓量的情况。

历史老照片:老上海商埠码头一景
historical photo: part of the wharf in Shanghai

历史老照片:20世纪30年代的"永兴仓库"
historical photo: Yung Shing Warehouse in the 1930s

No. 61 Yangshupu Road project used to be Yung Shing Warehouse, a six-storey building of 22.6 meters high, covering a land area of 6,227 square meters and a gross floor area of 37,362 square meters. This large building is just like a solid castle, standing in the shipping hub area of Yangshupu Road, Huimin Road and Huishan Wharf, enjoying a superior location.

Yung Shing Warehouse is one of the most famous warehouses in Shanghai. In 1929, A. Corrit Co. built a four-storey reinforced concrete warehouse on Baikal Road (now No. 36 Huimin Road). It lies at the junction of Hongkou District and Yangpu District, the most concentrated area of early industry, trade and shipping in Shanghai. As a result, the demand for warehousing was often oversupply.

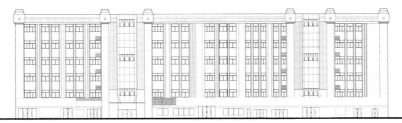

杨树浦路61号立面改造设计图
renovation design sketch of the appearance of No. 61 Yangshupu Road

为了满足进货需求，扩大仓储业务，1935年康益洋行又以衔接的方式在南侧再建一幢六层钢筋混凝土仓库，使仓库总面积达3427平方米，为同类仓库之最。

整栋建筑中最具特色的当属享有"远东第一梯"美誉的两部由美国奥的斯公司（OTIS）设计制造的8吨载重货运电梯。卡车可驶入电梯轿厢，经运载货车自如到达各楼层，最早实现了"门到门"的货运方式。它方便货运装卸，能减少重复人力劳作，节约劳力，故而享有"远东第一流设备仓库"之美誉。

历史老照片：正在运载货物的"远东第一梯"能直接将货物运送到各楼层
historical photo: "No.1 Elevator in the Far East" was transporting the goods to each floor

In order to meet the demand of purchase and expand warehousing, A. Corrit Co. built another six-storey reinforced concrete warehouse on the south side in 1935 in a way of connection, with the gross area of 3,427 square meters, making it is the largest of its kind.
The two 8-ton freight elevators designed by the U.S. elevator company OTIS, which were hailed as the "No.1 Elevator in the Far East", are the most distinctive. The truck could drive into the elevator car and reach each floor freely, which is the earliest door-to-door freight transportation. Yung Shing Warehouse was praised as the "First-Class Equipment Warehouse in the Far East" for its convenient cargo loading and unloading, which could reduce repetitive labor and save labor.

在战乱中，"永兴仓库"几易其主，库名也随之更改为"汇山堆栈""扬子仓库"，1946年归属上海永兴地产股份有限公司后，才恢复"永兴仓库"的名称。1949年5月28日，永兴仓库为上海商业储运公司使用。

曾更名为"扬子仓库"的历史物证
historical physical evidence of rename into Yangtsze Warehouse

目前展示的"远东第一梯"
Current "No.1 Elevator in the Far East"

Yung Shing Warehouse had a succession of owners in the war, and it was renamed Huishan Godown and Yangtsze Warehouse as well. After being incorporated into Shanghai Yung Shing Real Estate Co., Ltd. in 1946, the name of Yung Shing Warehouse was resumed. On May 28, 1949, Yung Shing Warehouse was used by Shanghai Commercial Storage and Transportation Co., Ltd.

唤醒
RENAISSANCE

150年前的故事早已泯灭在历史的洪流中，这条承载着百年工业兴衰的道路也在城市发展的进程中被遗忘：道路基础设施老化，风貌建筑群在历史的尘埃里被掩埋……随着城市更新的时代浪潮袭来，以百联资控为代表的企业对这些工业老建筑进行保护性开发，老建筑的更新将重构这条百年工业之路的肌理与气质；百联资控也将在打造北外滩百联金融产业办公集群，提升提篮桥文化历史风貌区整体形象的品牌增值工程中打磨出一张魅力名片。

The story 150 years ago had long been lost in the history. This road bearing the rise and fall of industry for a hundred years has also been forgotten in the process of urban development: the infrastructure is aging, the buildings with historical features are being buried in the dust of history… With the trend of urban renewal in the new era, enterprises represented by Bailian Asset Holding are engaged in protective development on these old industrial buildings, and the renewal of old buildings will in turn reconstruct the texture and temperament of this century-old industrial road; Bailian Asset Holding will also polish an attractive business card in the brand value-added project of creating Bailian financial industry office cluster in the North Bund and improving the overall image of the Tilanqiao zone with cultural and historical features.

气势宏大的建筑之韵

　　这座建筑体量庞大的仓库，在20世纪30年代是国内罕见的具有巴洛克式风格的工业建筑。这种风格在17世纪至18世纪中叶，流行于欧洲和拉丁美洲，在建筑上表现为外形自由，追求动态，喜好富丽的雕刻装饰，具有强烈的色彩，平面布局上常用穿插的曲面和椭圆形空间。

　　该仓库的建筑立面采用了夸张的处理手法，无论是巨大的柱式还是细腻而繁复的雕饰，错综凹凸的曲线、曲面和室内外空间的光影变幻，具有强烈的透视效果。这也是西方古典主义建筑师往往偏爱巴洛克风格的原因之一。

Magnificent Construction Style

This huge warehouse was a rare industrial building in Baroque style in the 1930s. The Baroque style was popular in Europe and Latin America from the 17th century to the mid-18th century. It is characterized by free appearance, dynamic pursuit, rich carving decorations, strong color, and curved surfaces and oval space interspersed in the plane layout.
The facade is processed with exaggeration, evidenced by the huge columns or delicate and complex carvings. Intricate and convex curves, curved surfaces and the light and shadow changes in the indoor and outdoor space show a strong perspective effect. This is one of the reasons why Western classical architects prefer Baroque style.

◀ 杨树浦路61号建筑南立面实景照
photo of the south facade of No. 61 Yangshupu Road

杨树浦路61号建筑南立面实景照
photo of the south facade of No. 61 Yangshupu Road

　　永兴仓库的建筑平面为矩形，立面构图中心为方正平拱窗，两侧有高大的壁柱，顶层女儿墙出檐处有方形齿状小装饰，细部很别致。沿街立面由三个单体建筑通至楼层平台相互连接而成，立面垂直简洁，柱间为坚固钢制横窗，端正典雅。

The building plan of Yung Shing Warehouse is rectangular. The center of the facade composition is square flat arched windows, on both sides of which are high pilasters. The eaves of the top parapet are decorated with small square dentils, featuring unique details. The facade along the road is composed of three single buildings connected to the floor podium base. The facade is vertical and simple, and between the columns are solid steel transoms, upright and elegant.

百联资控团队在杨树浦路61号现场考察
on-site survey of No. 61 Yangshupu Road by Bailian Asset Holding team

　　楼内排列有400多根六角形"梅花柱"，柱上栓有六角形柱帽，楼板为无梁砼板，由钢混浇筑而成。由于丹麦设计师的匠心，充分考虑到仓库内采光、通风、卫生、消防等设施，使建筑构思精致，设计巧妙，房屋气势雄伟，规模宏大，整体建筑带有欧式韵味，是上海近代优秀仓库建筑之一。

Inside the building, there are more than 400 hexagonal plum blossom columns bolted with hexagonal column caps. The floor slab is girderless using reinforced concrete. Thanks to the ingenuity of Danish designers, the full account of facilities including lighting, ventilation, sanitation, and fire fighting contributes to the exquisite concept, ingenious design, and majestic style with some European elements. Yung Shing Warehouse has therefore become one of the excellent warehouse buildings in modern Shanghai.

由于主楼后于副楼建造,因此延续了副楼的流线布局,平面上分为三个仓储单元,西北-东南方向布局,之间沿建筑设中庭,底层为半室内架空车行道,上层开敞,南北各以一部楼梯连接。

Built later than the wing building, the main building has continued the streamline layout of the wing building. It is divided into three warehousing sections on the plane. The layout is from the northwest to the southeast, with an atrium set between them along the building. The ground floor is a semi indoor overhead roadway, with the upper part open, and the north and south are connected by a staircase.

项目改造室内中庭,极大提升采光和通风
the renovated indoor atrium greatly enhances the lighting and ventilation

项目平面布局图
plan of the project

项目在2007年经历过一次改造。主楼原建筑单元之间的室外空间被改造为室内中庭,设有天窗;建筑内部空间格局也进行了改动,原仓库建筑大空间增加了隔断,调整为单元式的办公空间。

The project was renovated in 2007. The outdoor space between the original units of the main building has been renovated into an indoor atrium with skylights; the internal space pattern has also been changed: partitions are added to adjust the large space into unit-based office space.

躺进旧时光里的重生

城市发展是一种积累，呈现出"底片叠加"的效果，它会反映出过去的影像，而不该像照片覆盖那样将过去一笔抹杀。老建筑恰恰是凝固过去影像、文化的载体，所以要根据合理的价值评估，予以适当保留和改造。

程大利女士介绍："大规模的开发、推倒重来的模式基本上已经走到头了，我们提出的方案是用节点规划的模式，通过这种方式，去更新、替代一些小的节点，并不止于物理上的更新，更多的可能是软件或是内容的植入，并从这些节点辐射周边。"

2019年7月5日上午，由虹口区文物部门在杨树浦路61号会议室组织召开了"杨树浦路61号装修项目"方案评审会，参会专家在听取方案后，经认真讨论形成了专家意见。

Rebirth in the Old Days

Urban development is a kind of accumulation, showing the effect of "film superimposition". It reflects the past images, rather than obliterating the past like photo coverage. Old buildings are the carrier of solidifying the past images and culture, so they should be properly preserved and renovated according to the reasonable value assessment.

According to Ms. Cheng Dali, "The mode of large-scale construction and demolition has basically come to an end. The proposal we put forward is to apply the mode of node planning to update and replace some small nodes, including not the physical update, but also the implantation of soft decorations or content, and finally to radiate the surrounding from these nodes around."

In the morning of July 5, 2019, the Cultural Relics Department of Hongkou District organized and held the review meeting of the Decoration Plan in the conference room of No.61 Yangshupu Road. After listening to the scheme, the present experts formed their opinions upon careful discussion.

透过玻璃幕墙看杨树浦路61号
viewing No. 61 Yangshupu Road through glass curtain wall

唤醒
RENAISSANCE

项目本着真实性和可逆性原则，进行全面装修改造与空间升级。重点是重现历史风貌，突出建筑风格特色，提高室内外品质，明确特色业态，活化室内空间。同时，还将整饬外部环境和交通组织，扩容停车位，更新设施设备，替换老旧系统，提升大楼运能与系统效率。

建筑外观修缮工程的主要内容是针对外立面主体采用黄色水泥砂浆进行粉刷，尽量恢复历史原貌。波纹装饰线脚采用GRC（Glass fiber Reinforced Concrete 的缩写，中文名称是玻璃纤维增强混凝土）预制干挂，外面用黄色水泥粉刷，且比例分隔力求按照历史图纸和照片复原。外门窗恢复钢门窗型制，采用黑色油漆。窗台、勒脚、檐口和窗间装饰线条都采用和黄色水泥色调接近的水刷石饰面。

项目南立面改造效果图
renovation design sketch of the south facade

In line with the principle of authenticity and reversibility, comprehensive renovation and space upgrading will be done, with focus on reappearing the historical features, highlighting the style, improving both the indoor and outdoor quality, clarifying the characteristic business forms, and activating the indoor space. Efforts will be made to adjust the external environment and traffic organization, expand the parking space, update the facilities and equipment, and replace the old systems, in order to escalate the building capacity and system efficiency.

The facades have mainly been painted with yellow cement mortar as an attempt to restore the original appearance. The corrugated decorative architraves use Glass fiber Reinforced Concrete (GRC) for prefabricated dry hanging, painted with yellow cement, and the proportional separation are also to be restored according to the historical drawings and photos. The exterior door windows will be made of steel, and painted in black. The windowsills, plinths, cornices and decorative lines between windows will be decorated with washed granolithic plaster veneer close to the color of yellow cement.

项目北立面改造效果图
renovation design sketch of the north facade

唤醒
RENAISSANCE

二层中庭改造效果图　renovation design sketch of the atrium on the second floor

室内空间保护了原无梁楼盖结构体系，原先的办公空间被重新划分，用于金融产业办公、特色商业和艺术空间。办公空间采取简约的设计格调，体现原有建筑质感。商业空间内增加共享空间，方便人与人之间的交流，在空间上增加视觉层次感。

　　本项目的设计风格主题定位为"与时光的融合蜕变"。设计师认为，原建筑在时间轨迹的冲刷下，其巴洛克建筑风格有所侵蚀，所以现在除了需要保持原有建筑质感以外，还要在设计中加入现代艺术手法，在过去与现在的历史轨迹碰撞中再赋予装饰面新的使命和活力，引起人们探询历史的兴趣。室内装饰色调以白色、高级灰为主基调，加入"光"元素来解决建筑本身采光性的短板，突出"历史"面和结构柱该有的形态和质感。

Indoor space protects the original girderless floor structure system. The original office space is re-divided into the financial industry office space, characteristic commercial space and art space. The office space adopts a simple design style to reflect the original texture. In the commercial space, shared space will be added to facilitate the communication between people, and increase the sense of visual hierarchy.

The design style theme of this project is "integration and transformation with time". According to the designers, the Baroque style of the original building has been eroded over time, so in addition to maintaining its original texture, modern art techniques should be added to endow a new mission and vitality of the veneer in the collision between the past and the present, in order to arouse people's interest in exploring the history. The interior decoration tone will be mainly white and high-grade gray, with the "light" element added to solve the insufficient lighting and highlight the form and texture of the "historical" nature and structural columns.

项目改造剖面效果图
renovation design sketch of section of the project

项目手记

由于老仓库的空间格局和形态轮廓已形成历史记忆，尊重历史、保护原建筑主要的内部空间格局和建筑群体的造型成为工作重点。项目通过价值评估，设计师以激活老仓库为目标，采用了部分保护、部分更新的策略，在保证新功能使用的同时，兼顾保存历史记忆，使城市历史减少断层，延续发展。

未来，走进杨树浦路61号，你会惊喜地看到，这是一个多么独特的大楼，新装修的大堂、走廊宽敞明亮，载重8吨的"远东第一梯"作为观光电梯被长久地保留在这座大楼内部，见证繁华与沧桑。而新型的办公电梯快捷地匀速上下，各办公空间秩序井然。报关、国际贸易、国际物流、货物代运等行业新型企业比比皆是。杨树浦路61号将用自己独一无二的视角去见证杨树浦路这片地段的盛世繁华。

Project Notes

The spatial pattern and shape outline of this old warehouse have formed into a historical memory, so the renovation focus is on protecting the main internal spatial pattern and the shape of this original building on the premise of respecting the history. Upon value assessment, with the aim of activating this old warehouse, the designers have adopted the strategy of partial protection and partial renewal to preserve the historical memory while ensuring new functions, thus reducing the fault of urban history and continuing urban development.

In the future, when you walk into No. 61 Yangshupu Road, you will be surprised to find how a unique building it is. The lobby will be newly decorated. The corridor is spacious and bright. The No.1 Elevator in the Far East with a load of 8 tons will be kept in the building for long as a sightseeing elevator to witness the development of Shanghai. The new-type office elevator can go quickly and evenly up and down. Offices are in order. Industries such as customs declaration, international trade, international logistics, freight forwarding will be full of new-type enterprises. No. 61 Yangshupu Road will witness the prosperity of the road from its unique perspective.

杨树浦路61号现状特写
close-up photo of the current No. 61 Yangshupu Road

第七章
未来绽放：马登仓库（杨树浦路147号）
—— 基于场所精神的老建筑改造

Chapter Seven
A Bright Future: Pier Marden (No. 147 Yangshupu Road)
– *Renovation of the Old Building Based on the Spirit of Place*

马登仓库大楼外观实景照
photo of the facade of Pier Marden

  北外滩的航运历史沉淀至今已有160多年。1845年，英商东印度公司在徐家滩（今东大名路、高阳路）一带建造了驳船码头；1868年，英商蓝烟囱轮船公司大型远洋轮船"鸭家门号"首次停泊在虹口码头，自此开启了上海对外航运的大门。

  如今，位于北外滩的上海港国际客运中心集合了国际客运码头、港务大楼、写字楼以及艺术画廊、音乐文化中心等，邮轮经济成为上海国际航运中心建设的新亮点，北外滩将成为国际邮轮公司母港的最佳选择，它的积累和辐射功能，在未来几年将远远超出我们的想象。

  马登仓库创意园区的改造是一个将历史建筑保育再活化的典型案例。新设计保留了西式工业建筑的原有外观，并通过引入适应城市新需求的功能，延长了建筑的使用寿命，打造出一个全新的集聚国内外创意类、设计类企业的创意园区。

  从清晨到日暮，大楼内的光影变化迷人而丰富。中西合璧、新旧并存，动感时尚与老工业建筑风格相辅相融。而北外滩蒸蒸日上的发展，产业升级带动的经济活力、人才聚集高地效益等，也为老建筑投射了更动人的剪影。

The North Bund has a shipping history of more than 160 years. In 1845, the British East India Company built a barge pier in Xujiatan (now East Daming Road and Gaoyang Road); in 1868, the large ocean-going vessel Yajiamen of UK Blue Funnel Line was first berthed at Hongkou Wharf, opening the door of Shanghai's foreign shipping.

Today, Shanghai Port International Cruise Terminal located in the North Bund has pooled international passenger terminals, port buildings, office buildings, art galleries, music and culture centers, etc., making the cruise economy a new highlight amid its construction. The North Bund will become an optimal choice for the home ports of international cruise companies. Its function of accumulation and radiation will far exceed our expectations in the next few years.

The renovation of Pier Marden Creative Park is a typical case of reactivating a preserved historical building. By retaining the original appearance of this western style industrial building, and extending its service life after introducing the functions that adapt to the new needs of the city, the new design has built a new creative park pooling creative and design enterprises at home and abroad.

The charming rich light and shadow changes continue from dawn to dusk. Chinese and western styles are mixed. Old and new decorations coexist. Fashion and historical elements complement each other. The booming development, the economic vitality driven by industrial upgrading, and the benefits brought by talent pooling in the North Bund are also casting a more moving silhouette on the old buildings.

类叠历史——过去未去

在杨树浦路上，各式各样的老建筑、现代民居和新建高楼交错密布，高高低低地勾勒出无限繁华的剪影。风格迥异的老建筑都是颇有年头的存在，有的仍在使用，有的转租，有的闲置。时间藏在砖缝里，历史游走于窗框前，每幢房子都有故事，让人不经询问，有哪些故事我们未曾知道。

Overlap on the History: The Past Has Not Gone

On Yangshupu Road, there are a variety of old buildings, modern residences and new high-rises crisscrossed to silhouette Shanghai. These time-honored old buildings in different styles are in use, or rented, or idle. Time is hidden in the cracks of bricks. History travels in front of the window frames. Every old building has a story, and there may be some we have never heard.

大楼外墙上的文保单位铭牌
nameplate of historical and cultural protected site on the appearance

杨树浦路147号立面及石柱

appearance and stone pillars of No. 147 Yangshupu Road

第七章 CHAPTER SEVEN

走到绿树成荫、宽畅空阔的杨树浦路一侧，杨树浦路147-155号马登仓库与61号永兴仓库仅仅隔了一个晋阳里（上海里弄式民居）。这栋红色老建筑很吸引人，用手机扫描墙上的"马登仓库旧址"铭牌二维码，一段关于它的历史就呈现在眼前。

高度倚仗水运的年代，杨树浦路是各路客商登陆上海的第一站，英商茂泰洋行在这里就有很多泊位和仓库。茂泰洋行由原江海关三等总巡马登（G.E.Marden）等人创办于1925年，公司总部设在汉口路125号，经营报关、运输、驳运、拖船、仓储等业务。随着业务发展需要，茂泰洋行聘请英籍建筑师马登于1929年设计了这幢现代式建筑风格仓库，1930年建成使用，并以他的名字命名，故称"马登仓库"。

马登仓库大楼外观实景照
photo of the facade of Pier Marden

Walking by the broad Yangshupu Road planted with trees, Pier Marden at No. 147-155 is only away from Yung Shing Warehouse at No. 61 across Jinyang Alley (an alley-style residence in Shanghai). This old red building is quite attractive. Scan the QR code of the nameplate on the wall with your mobile phone, its history will be presented in front of you.
IIn the era of high reliance on water transport, Yangshupu Road was the first stop for merchants from all walks of life to land in Shanghai. There were many berths and warehouses of Marden & Co., Ltd. Founded by the former Third-class Boat Officer G.E. Marden of Jiangnan Custom House in 1925, Marden & Co., Ltd. was headquartered at No.125 Hankou Road, specializing in customs clearance, transport, barging, tugboat, warehousing, etc. As the business grew, Marden & Co., Ltd. hired British architect Marden to design this warehouse in modern style in 1929, which was completed and used in 1930 and named after him.

杨树浦路147号鸟瞰图
aerial view of No. 147 Yangshupu Road

杨树浦路147号的红砖立面
red-brick appearance of No. 147 Yangshupu Road

唤醒
RENAISSANCE

　　马登仓库是一幢高22.4米的六层楼仓库建筑，占地面积3242平方米，建筑面积16212平方米。钢筋混凝土结构（局部砖混结构）的仓库，宛如一座坚固的城堡，耸立在杨树浦路、秦皇岛渡口及汇山码头之间的要冲地区。

　　马登仓库是上海目前保留较为完整的西式工业建筑，是20世纪30年代国内早期的现代主义风格的工业建筑。

　　此建筑风格产生于19世纪后期，成熟于20世纪20年代末，在20世纪30—40年代风行全世界。当时建筑师为了摆脱传统建筑形式束缚，需要大胆地创造适应于工业化社会需求的新建筑。因此在工业建筑类型，如工厂厂房、码头仓库中广泛实践。设计师抛弃了各种仿古典建筑的繁琐装饰，采用简洁、实用、平面自由、讲究功能的方式和理念设计作品，马登仓库乃是其中一例。

第七章
CHAPTER SEVEN

Pier Marden is a six-storey warehouse building with a height of 22.4 meters, covering a land area of 3,242 square meters and a gross floor area of 16212 square meters. The warehouse with reinforced concrete structure (partial brick-concrete structure) is like a solid castle, standing in the key area among Yangshupu Road, Qinhuangdao Ferry and Huishan Wharf.

Pier Marden is a well-preserved industrial building in Western style in Shanghai. It was also an industrial building in early modernism style in China in the 1930s.

This style originated in the late 19th century, matured in the late 1920s, and became popular all over the world in the 1930s–1940s. In order to get rid of the shackles of traditional architectural forms at that time, architects needed to boldly create new buildings adapted to the needs of industrial society. Therefore, it was widely applied in industrial building types, such as factory plants and wharf warehouses. Designers abandoned all kinds of complicated decorations in classical style and designed in a simple, practical, free plane and functional way. Pier Marden was one of them.

马登仓库外墙实景照
photo of the exterior wall of Pier Marden

简洁现代的立面现状
current simple and modern appearance

杨树浦路147号的仓库结构
Warehouse structure of No. 147 Yangshupu Road

杨树浦路147号的红砖立面
red-brick appearance of No. 147 Yangshupu Road

建筑西立面
west facade of the building

马登仓库的建筑呈立方体造型，建筑沿街立面简洁，柱间为钢制横窗，端庄典雅。库内结构设置巨大的巴西利卡式钢筋混凝土水泥梁架，排列十分雄伟。建筑师还充分考虑到采光、通风、卫生、消防等设施布局，构思精致，设计巧妙。房屋气势宏伟，规模庞大，整体带有欧洲风格，是近代上海优秀仓库建筑之一。

The cubic building of Pier Marden features simple facades along the road, and between the columns are solid steel transoms, upright and elegant. There is a Basilica reinforced concrete beam frame, arranged in a magnificent way. With full account of the layout of facilities such as lighting, ventilation, sanitation, and firefighting, the architect delivered exquisite conception and ingenious design. This large magnificent building in European style is one of the outstanding warehouse buildings in modern Shanghai.

忠于文脉——传承重构

相关资料显示,至1953年马登仓库仍作为当时茂泰商业公司的仓库使用,1963年收归国有,属于上海市国营商业储运公司,命名为提篮桥仓库。

原仓库内部为大空间框架结构,无特殊建筑装饰;2007年本建筑作为创意园区进行了整体改造,室内各层采用石膏板墙等进行小开间分隔作为办公使用,并进行吊顶处理。

Loyalty to the Context: Inheritance and Reconstruction

According to relevant data, in 1953, Pier Marden was still used as the warehouse of Marden & Co. It was nationalized in 1963, belonging to Shanghai Commercial Warehousing & Transportation Co., Ltd., and named Tilanqiao Warehouse.

The interior of the original warehouse was a large space frame structure without special building decorations; but in 2007, the building was renovated as a creative park as a whole, and gypsum board walls were separated for office use with furred ceiling.

马登仓库创意园区入口　　entrance of Pier Marden Creative Park

2013年，马登仓库被公布为虹口区文物保护单位，但项目历次修缮对建筑干预较多，历史原状已保留不多。本次修缮工程以不改变原建筑主体承重结构为原则，并保留下外立面原有装饰线脚及装饰面层材料。室内楼梯是体现和反映工业遗产类文物建筑特征的重点保护部位，也将进行复原修缮。

为明确建筑物外立面完损情况，文物勘察单位进行了现场勘察检测，发现房屋外墙主要存在装饰线开裂、外墙多处残留空调支架、底层花岗岩装饰局部缺损、局部室内墙体平顶渗漏水等情况。

柱脚花岗岩缺损
defect of the bottom granite decorations

装饰线脚开裂
cracks of surbase

檐口局部损坏
partial defect of the eaves

局部墙面渗漏水，粉刷起皮脱落
partial leakage of water from walls and peeling paint on walls

外墙残留空调支架
disorderly air conditioner brackets

局部天沟渗水
partial leakage of water in ceiling channel

文物勘察单位勘测结果　investigation result by a cultural relics investigation entity

In 2013, Pier Marden was announced as a historical and cultural protected site in Hongkou District. Unfortunately, previous renovations interfered much with the building, leaving little historical original. In line with the principle of not changing the -bearing structure of the original building, this renovation will retain the original decorative architraves and surface materials on the facade. As key protected parts reflecting the characteristics of industrial heritage cultural relics, the indoor stairs will also be restored.

In order to make clear the damage extent of the facade, a cultural relics investigation entity carried out on-site investigation and detection, finding the cracks of surbase, the residual air-conditioning support in several parts, the partial defect of the bottom granite decorations, the partial leakage of the flattop, etc.

因此，在本次修缮工程中，外立面损坏部位采用同类同色同质的面层材料，以相同施工工艺进行修缮。对立面上空置的空调支架等无用的附属构件进行拆除后并对外墙进行修补，附属在外墙的电线等进行整理、归并。

建筑南立面 south facade of the building

Therefore, in this project, the damaged parts of the facade will be repaired with surface materials of the same color and quality and in the same construction process. Upon removal of useless auxiliary components such as the air conditioning bracket on the facade, the exterior wall will be repaired, and the wires attached to it will be arranged and lumped together.

檐口损坏修复，重新涂料
repair the eaves and repaint

外墙仿红砖面砖损坏处
defect of red ashlar bricks on exterior wall

用相应颜色同质面砖修复
repair the exterior wall with red ashlar bricks

窗台线损坏修缮
repair the window molding

花岗岩柱脚局部损坏修缮
repair the bottom granite decorations

建筑南立面修缮方案示意图　repair design sketch of the south facade

第七章
CHAPTER SEVEN

室内两部楼梯是原仓库的主要竖向交通部位，原设计充分考虑了当时仓库人工搬运这一特殊需求，踏步面宽、坡度平缓，设计大气、合理，具有一定的建筑研究及保护价值，也是本项目的重点保护部位。目前现状楼梯周边墙体粉刷起壳脱落，踏步漆面磨损，栏板扶手等存在不同程度的破损。

在本项目修缮工程中，地面饰面材料采用水泥砂浆地坪或类似水泥的自流平装饰面层，扶手栏杆采用原面层装饰材料修复方法。同时为满足消防疏散的要求，仅对该楼梯进行封闭式改造，整体楼梯风格不变。

The two indoor stairs are the main vertical traffic parts of the original warehouse. Upon full consideration of the special demand for manual handling at that time, the steps were designed wide with a gentle slope in a magnificent and reasonable style, which have certain research and protection value. They are also a key protected part of this project. The wall around the stairs have peeled off, the paint surface of steps is worn, and the handrails of the breast board are damaged to varying degrees.

楼梯修缮前现状照片
photo of the current staircase before repair

In this project, the cement mortar floor or self-leveling decorative surface similar to cement will be used for overcoating, and the handrails will be repaired with original surface decorative materials. In order to meet the demand for fire evacuation, closed renovation will be done to the stairs, without changing the overall style.

建筑外观修缮效果图
design sketch of appearance after repair

项目大堂改造效果图　renovation design sketch of the lobby

　　室内设计策略是保留整体空间格局及反映原建筑历史特征的部件,与全新设计的功能空间并置,使老仓库既有新时代气息又有旧建筑韵味,达到历史与现实融合并存。

　　室内空间用大块的白墙和玻璃来取代原有的纷杂凌乱,以此淡化多个承重梁柱把空间无形切割的碎片感,使整体更通透敞阔。地面规整的方形图案大理石花砖铺陈出一种既古典又现代的美感,让人一走进来就有空间切换的全新感受。

　　同时,大面积落地窗为办公空间提供充足的采光条件,暖暖的阳光无遮挡地被"请"进室内,配以大型绿植点缀其间,营造出绿意盎然的灵动氛围,为创意办公用户带来更多灵感体验。

办公空间改造效果图　renovation design sketch of the office room

The interior design strategy is to keep the overall spatial pattern and the components that reflect the historical characteristics of the original building, and to juxtapose with the newly designed functional space, so that the warehouse can have both the flavor of the new era and the charm of an old building, and achieve the integration of history and reality.

Large white walls and glass will replace the original clutter to weaken the sense of spatial fragments cut invisibly by multiple bearing beam columns, making the interior space more spacious and open. The regular square pattern marble tiles on the ground show the beauty with a mix of classical and modern elements, offering people a new feeling of space switching the moment they enter.

In addition, the large floor-to-ceiling windows provide enough lighting conditions for the office space, coupled with large-scale green plants, creating a green and dynamic atmosphere to bring more inspiration experience for creative office users.

唤醒
RENAISSANCE

连贯的空间与灵活的家具,使场地能够适应办公、展览、放映或聚会的需要。
Continuous space and flexible furniture enable the venue to meet the needs including office, exhibition, projection, and gathering.

创新空间——未来已来

　　未来，虹口将打好产业发展、功能提升、环境营造这套"组合拳"，力争把北外滩打造成为卓越全球城市的顶级中央活动区，展现城市文化特质和精神风貌的世界级会客厅。

　　目前，北外滩集聚了4000多家航运企业，1000多家金融企业。老旧的仓库正变身为并购基金和航运、科创企业的集聚地，绿色技术产业在这里生根。

　　杨树浦路147号将通过提供灵活租赁产品、拎包入住的环境与共享服务（如前台、门禁、社区等），打破传统，释放办公企业成本压力，为客户创造价值。

　　可以预见，未来这里将成为激发灵感的创作场所，是体验交流、共同成长的舞台，是企业品牌的理想之地。这里拥有现代化的创作工作室，公共的展示区域，并定期举行行业发布会，邀请艺术大师进行讲座和交流，在这里艺术智慧的火花将不断被点燃。

Innovation Space: The Future Has Come

In the future, Hongkou will make a good combination of industrial development, functional improvement and environmental construction, striving to build the North Bund into a top central activity area in this outstanding global city, and a world-class reception hall that displays the cultural characteristics and spiritual outlook of Shanghai.

There are more than 4,000 shipping enterprises and 1,000 financial enterprises in the North Bund. The old warehouse is also gathering enterprises engaged in M&A funds, shipping and science and technology, where the green technology industry takes root.

By offering flexible rental products, an environment available for immediate occupancy, and shared services (e.g. reception desk, entrance guard, and community), the No.147 yangshupu road will break the tradition, release the cost pressure of office enterprises, and create value for customers.

Predictably, the park will become a creative venue for sparking inspiration, a stage for experience exchanges and common growth, and an ideal place for corporate brands. There will be full of modern creative studios and public exhibition areas. Press conferences will be held regularly. Art masters will be invited to give lectures and exchanges. The spark of art wisdom will be continuously ignited here.

程大利女士在杨树浦路147号项目现场考察
on-site survey of No. 147 Yangshupu Road Project by Ms. Cheng Dali

项目手记

不同时期留存下来的街道、老仓库、老厂房是城市经济发展的印记，反映着特定时代的产业风向。这些宝贵的历史遗迹，成为"最忠实"的城市地域文化的见证者和讲述者。

作为城市更新的参与者和探索者，百联资控从市场和项目出发摸索出一种温和式城市更新思路。城市记忆场所的留存并非一定要完整无缺，很多承载着记忆的地方都只是将其最具特色的部分保存下来。这些老建筑再利用，都是在外形结构不变的情况下，对内部加以适度改造，从而让原本失意的老建筑重新进入社会生活，服务于新时期的发展空间需要。

Project Notes

The extant roads, old warehouses and old factories built in different periods are the marks of urban economic development, reflecting the industrial trend of a specific era. These precious historical sites have become the most faithful witnesses and narrators of urban regional culture.

As a participant and explorer of urban renewal, Bailian Asset Holding has pioneered a moderate urban renewal idea from the perspective of the market and projects. It is not necessary to keep the sites bearing urban memory completely intact. In many sites bearing urban memory, only the most characteristic parts have been preserved. The reuse of old buildings is to appropriately renovate the interior on the premise of not changing the external structure, so that the original frustrated old buildings can re-enter the social life and serve the development space needs of the new era.

* 注：本章节部分历史资料原引自薛顺生《杨树浦路上老仓库——原马登仓库与"61号老栈房"》。
*Note: Part of historical materials in this chapter refer to the *Old Warehouses on Yangshupu Road: Former Pier Marden and No. 61 Warehouse* by Xue Shunsheng.

杨树浦路147号保留着原有的红砖立面与石墙
retained red-brick appearance and stone wall of No. 147 Yangshupu Road

第七章 CHAPTER SEVEN

第三部分 PART 3

🍀 莘荟 社区商业产品线——融合创新的活力生活舞台
【从"厂"变"场"的区域升级】

🍀 莘荟 Community Commercial Project Line: A Dynamic Life Stage of Integrated Innovation
Regional Upgrading from an Old Factory to a Life Field

唤醒
RENAISSANCE

《礼记·大学》有云："苟日新，日日新，又日新。"

城市的有机更新就是上海"日日新"的主要成长方式，也是一种持续和创新的责任和使命，展现了"海纳百川、追求卓越、开明睿智、大气谦和"的上海城市精神，如同交响乐一般，不停地律动、跳跃、旋转、激扬。

在《上海市城市总体规划（2017—2035年）》报告中，政府提出要以提高城市活力和品质为目标，积极探索渐进式、可持续的有机更新模式，以存量用地的更新利用来满足城市未来发展的空间需求，同时做好城市文化的保护与传承，倒逼土地利用方式由外延粗放式扩张向内涵式效益的提升转变，促进空间利用向集约紧凑、功能复合、低碳高效转变。

我们畅想，2035年的上海，"建筑是可以阅读的，街区是适合漫步的，公园是最宜休憩的，市民是尊法诚信文明的，城市始终是有温度的。"

那么，构成城市温度基底的一个个街区，怎能在这一轮城市更新的浪潮中失语？那些潜藏在街区、与社区比邻的老旧厂房，不该被时间封印，它们已然成为城市更新的重要部分。

旧厂房和仓库作为工业时代的产物，不仅记载了城市文明和技术发展的历史，更承载了人们对于特定年代的深刻记忆。历经时间的打磨和风雨的洗礼，流水线上机器的轰鸣作响离我们远去，这些旧厂房不再具备实用功能，逐渐被城市淘汰……一切归于平静的背后，新生命即将生根发芽。

According to the *Great Learning of the Book of Rites*, "If you could cleanse yourself today to have a fresh start in life, you can have a fresh start every day and insist on having it daily forever."

In a similar way, the organic renewal of Shanghai is a continuous and innovative responsibility and mission, showing the spirit of Shanghai of being inclusive, excellent, wise, and humble, like a symphony, rhythmical, jumping, revolving, and inspiring.

In the report of *Shanghai Master Plan (2017-2035)*, the government proposed to take improving urban vitality and quality as the goal, actively explore a progressive and sustainable organic renewal mode, meet the space demand of urban development in the future through the renewal and utilization of in-stock land, and do a good job in the protection and inheritance of urban culture, thus forcing the land use mode to shift from extensive expansion to the improvement of implicit benefits and to an intensive, compact, functional, low-carbon and efficient manner.

We hope that by 2035, Shanghai will become a city "where buildings are enlightening, streets are strolling-friendly and parks are enjoyable, while the citizens have the reputation of being law-abiding, credible and well-mannered. Shanghai, a city endowed with tenderness."

Then how can the blocks that form the base of urban temperature keep silent in this wave of urban renewal? Those old factories hidden in the blocks and adjacent to the communities should not be sealed by time. They have become an important part of urban renewal.

As a product of the industrial age, the old factories and warehouses not only record the history of urban civilization and technological development, but bear people's deep memory of a specific era. Over time, the roar of machines on the assembly line has been far away from us. The old workshops no longer have practical functions, but are being eliminated by Shanghai… Behind the tranquility, new life is about to take root.

旧工厂和仓库再利用对城市发展具有重要的意义，是现阶段国内城市更新、空间再生的一种重要手段。郑时龄院士说："城市遗产无疑包括大量工业历史的空间遗产，而工业遗存作为城市发展产业转型的旧痕迹，如同城市日记，有着独特的保存价值。"

但是，我们也发现不少老工厂改造后的场所和市民的联系非常微弱。这些地方似乎有一个无形的场，把圈子以外的人隔离开来。那里往往会成为某个特定人群的专属场所，并没有发挥出更大的社会效益。

建筑、街区是一个活的机体，要实现区域生活的真正升级，必须从街区"主人"的真正需求出发，使失落的城市场所重焕生机，让旧建筑成为有温度，有影响力，有凝聚力的新街区。

对此，百联资控以自身经验走出了一条从"厂"到"场"的区域升级、通过自身运营改变生活模式的探索之路。在保留历史记忆的同时，百联资控团队挖掘场所的个性，进行功能和建筑艺术上的重新诠释，提升地块的知名度与商业价值，使得往日沉睡的工业遗产成为充满活力的社区公共空间，以满足社区居民生活"最后一公里"的需求。以小而美、温馨精致的体验式场景消费为主打，融合了便利性的生活配套，从历史的折痕里展开当下生活的新场所、新舞台。

The reuse of old factories and warehouses is of great significance to urban development, and is an important means of urban renewal and space regeneration in China at this stage. "Undoubtedly, urban heritage contains a large number of spatial heritages of industrial history, and industrial heritage, as an old trace of urban development and industrial transformation, has a unique preservation value, just like urban diary," said the academician Zheng Shiling.
However, we also find that many of the renovated old factories have quite weak connections with the citizens. They seem to have an invisible field, separating people outside it. They often become exclusive to a specific group of people, and do not play a greater social benefit.
Buildings and blocks are a living organism. In order to realize the real upgrading of regional life, we must start from the real needs of the "owners" of the blocks. It is necessary to rejuvenate the lost urban sites, making the old buildings into new blocks with temperature, influence and cohesion.
Hence, Bailian Asset Holding has explored a road of regional upgrading from factories to sites with its experience to change the life mode through operation. While retaining the historical memory, Bailian Asset Holdings team excavates the individuality of the sites, reinterprets their functions and design art, and improves the popularity and commercial value of the plots, making the drowsy industrial heritage a vibrant community public space that meets the life need of the last kilometer of community residents. Focusing on the small but beautiful, warm and exquisite experiential scene consumption, coupled with the convenient life-supporting, the new sites and stages of current life will be shown through history.

唤醒
RENAISSANCE

第八章
活力舞台：北宝兴路624号
——立足于立体街区形态的活力社区

Chapter Eight
A Dynamic Stage: No. 624 North Baoxing Road
– *A Dynamic Community Based on Three-dimensional Block Form*

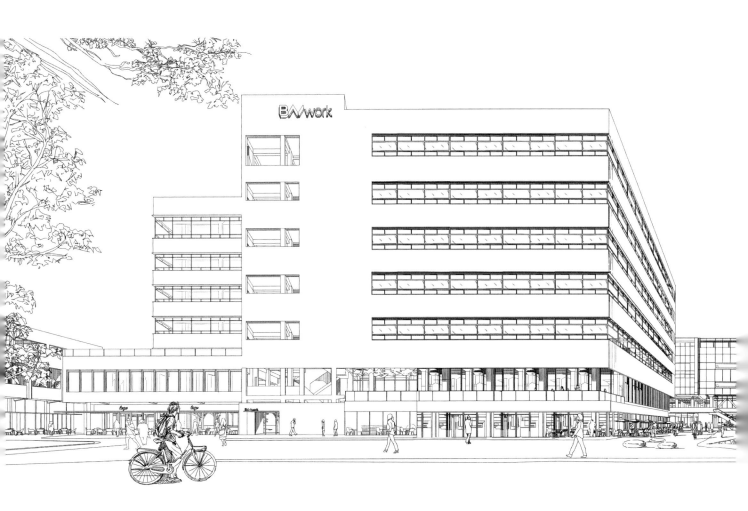

第八章
CHAPTER EIGHT

上海静安区北宝兴路624号，距上海大学延长路校区仅700米，一片安静的老冷链仓库厂房被三面居民区包围着，掩映在四周绿树浓荫里，也似乎被遗留在了旧时光里独自寂寥。这里曾经是上海联华生鲜食品加工配送中心，往日里井然有序的生产劳作场面随着工厂仓库的停工搬迁而戛然而止。

厂区内共10幢老建筑，这些老建筑不光承载了工人们的集体回忆，还蕴藏着过去热情洋溢的生活气息，更是一个时代的缩影。走过那个时代的人，一定会有特别的感受，每条道路、每个车间都可以牵出一段浓郁芬芳的时光记忆。

在这片厂房的四周，是一片密集的居民区，再往西过了南北高架则是北上海热闹的大宁商圈，那里矗立着大宁国际广场、大宁中心广场以及即将开业的大宁久光百货等大型购物中心，喧嚣的人流、车流，烘托着这片土地滚烫的价值。

如果说老旧厂房改造是历史留给城市建设的一张试卷，那么这一次，则是集百联资控同仁之智的最佳答卷。

通过百联资控的倾力打造，这片闲置的厂房将彻底"改头换面"，蜕变为面向社区消费者集生活集市、餐饮娱乐、儿童亲子、健身医美、运动休闲和社区服务特色商家，融便利、服务、体验、社交为一体的现代家庭全生活舞台。商业、办公、长租公寓的多元业态将为这片土地注入新的活力，为老厂房激发第二次生命力。

No. 624 North Baoxing Road in Jing' an District, is only 700 meters from the campus of Shanghai University on Yanchang Road. This quiet old cold-chain warehouse is surrounded by residential areas in three directions, hidden in the green trees, seemingly left alone in the old time. It used to be Shanghai Lianhua Fresh Food Processing & Distribution Center. The past orderly production and work scene stopped abruptly with the shutdown and relocation of the warehouse.

There are totally ten old buildings not only bearing the collective memory of workers, but containing the passionate vitality of the past and epitomize an era. People living in that era must have some special feelings. Every road and every workshop could lead to a rich memory of time.

There is a dense residential area around the factory, and across the South-north Elevated Highway westwards is the bustling Daning Business District in the north of Shanghai. There are large shopping centers including Daning International Business Plaza, Daning Central Square and the upcoming Daning Jiuguang Department Store. The bustling flow of people and cars highlights the top value of this land.

If the renovation of old factories is a test left by the history for urban construction, this project is the best answer of Bailian Asset Holding after pooling the wisdom of all the peers.

Through the efforts of Bailian Asset Holding, these idle factories will be completely renovated into a modern family whole-life stage that integrates convenience, services, experience and social intercourse for community consumers, including life market, catering and entertainment, parent-child activities, fitness and medical beauty, sports and leisure, and community services of featured merchants. Diversified business forms including commerce, office and long-term rental apartments will inject new vitality into the land and give the old factories a new life.

改造前航拍图 aerial photo before renovation

穿越历史 邂逅经典

老仓库的见证

北宝兴路624号地块原为上海联华生鲜食品加工配送中心有限公司，总占地面积21846平方米，总建筑面积44426.71平方米，由10幢大大小小的建筑错落组成，包括生产车间、冷库、配送场地、待发库、仓库（地下室）、办公楼、生活楼等。

这里曾一度是国内设备最先进、规模最大的生鲜食品加工配送中心。加工车间无异于一个"小小联合国"，来自中国、德国、美国、日本等世界各地的先进设备在厂房里各司其职，提供成熟、稳妥的生鲜食品加工工艺，多达1200种生鲜食品从这里源源不断地被生产加工出来，经由各大超市、生鲜卖场销售到全市老百姓的家庭餐桌上，大大提升了市民的餐桌幸福指数。在生产加工的同时，配送中心还从事水果、冷冻品以及南北货的配送任务。

Witness of the Old Warehouse

No. 624 North Baoxing Road used to be Shanghai Lianhua Fresh Food Processing & Distribution Center Co., Ltd., covering a land area of 21,846 square meters and a floor area of 44,426.71 square meters. It consists of ten buildings of different sizes, including production workshop, cold storage, distribution site, due-out room, warehouse (basement), office building, and living building.

It was once the largest fresh food processing and distribution center with the most advanced equipment in China. The processing workshop was just like a "small United Nations", in which advanced equipment from such countries as China, Germany, the United States, and Japan performed their duties and provided mature stable processing technology of fresh food. Up to 1200 kinds of fresh food were continuously processed from here, and sold to citizens through supermarkets and fresh stores, greatly improving the happiness index of citizens. The center was also engaged in the distribution of fruits, frozen products, and goods with origin of both northern and southern China.

原华联生鲜食品加工配送中心厂区门牌
nameplate of the former Shanghai Lianhua Fresh Food Processing & Distribution Center Co., Ltd.

原厂区和生产车间旧照
photos of the original factory and workshop

第八章
CHAPTER EIGHT

　　上海联华生鲜食品加工配送中心的建设可以说是城市民生工程的一枚闪亮勋章。在20世纪90年代末几年，上海市政府继"菜篮子工程""米袋子工程"之后，又加紧实施了"厨房工程"，其目的在于加快"厨房劳动的社会化"，使得千家万户从"买、汰、烧"繁重劳累的厨房劳动中解放出来，让居民吃得科学、吃得营养、吃得省时省力、吃得富于情趣。

　　由工厂生产的"厨房食品"通过"冷链"储运和绿色包装供应到市场上，市民选购后回家简单加热或烹煮即可解决一日三餐的问题。在那个外卖餐饮还不发达的年代，超市冷藏柜里的成品和半成品菜可称得上是都市"速食"解决方案中的"排头兵"了。

　　作为这一工程的重要服务配套，上海联华生鲜食品加工配送中心建立起了符合国际标准的、符合卫生标准的、具有一定加工规模的企业，其年生产能力可达20000吨，作为上海市的示范工厂向社会推广。毫无疑问，这项工程不仅解决了老百姓吃的问题，也为食品工业化改革、发展都市型食品工业创造了极大的发展机遇。

The construction of Shanghai Lianhua Fresh Food Processing & Distribution Center is a stunning urban livelihood project. In the late 1990s, following the "vegetable basket project" and "rice bag project", Shanghai municipal government stepped up the implementation of the "kitchen project", aiming to accelerate the "socialization of kitchen labor" and liberate thousands of households from heavy kitchen labor, so that residents can eat scientifically, nutritiously, happily, and in a time-saving and labor-saving way.

The "kitchen food" was supplied to the market through the cold chain storage and transportation and green packaging. People could buy them and simply heat or cook at home. In the era without developed order delivery, the finished and semi-finished dishes in the refrigerated cabinets of supermarkets were an optimal urban "fast food" solution.

As an important service support for this project, Shanghai Lianhua Fresh Food Processing & Distribution Center established an enterprise with certain processing scale that meet international standards and hygienic standards, with an annual production capacity of 20,000 tons. It was promoted to the society as a demonstration factory in Shanghai at that time. Undoubtedly, this project has not only addressed the problem of eat, but created great development opportunities for the reform of food industrialization and the development of urban food industry.

唤醒
RENAISSANCE

但是，随着时代滚滚车轮的不断前进，随着城区功能规划的改变，老厂房早已搬迁，曾经一派生机勃勃的地方失去了"眷顾"，被遗忘在蒙尘一隅。

如今，走进这片厂区，建筑外墙受潮生苔，粉刷涂料疏松脱落，墙面上藤蔓植物肆意生长攀爬，仿佛要把整栋建筑"吃"进历史的漩涡里去。寂静无人的厂区道路上杂草丛生，室外钢楼梯锈蚀严重，裸露的钢筋提醒着人们，这是一处无人问津的工业遗址，冰冷、没有温度。只有走进车间厂房，从遗留下来的制冷设备管道、锅炉设备还能让我们遐想一番当年热火朝天的生产景象。

改造前老旧的厂区内部　interior of the old factory before renovation

程大利女士带领团队在北宝兴路624号现场考察
on-site survey of No. 624 North Baoxing Road by the team led by Ms. Cheng Dali

庆幸的是，在百联资控的城市更新蓝图中，有它重要的一席之地。百联资控的六大产品线中，首当其冲的就是 &萃苓 社区购物中心。北宝兴路624号由于其良好的地理位置，周边的居民区密集等优质因素，被定位于全新活力社区型购物中心，这也将是 &萃苓 的首个旗舰项目。"15分钟社区生活圈"的改造愿景，就像一把开启美好生活的钥匙，和着活力而优雅的节拍，拥抱未来的大门。

However, with the progress of the times and the change of urban functional planning, old factories have been relocated. Their past refulgence has also been forgotten in the corner.
Today, the external wall is damp and mossy, with loose and peeled paint, and the vines on the wall grow and climb wantonly. There are many weeds on the road of this silent area, and the outdoor steel stairs are seriously rusted. The exposed steel bars remind people of its status as an industrial site left without anybody to care for it, cold, without any temperature. Only when we enter the workshop can we imagine the past busy production scene from the left cooling equipment pipes and boiler equipment.
Fortunately, this project is important in the blueprint of urban renewal of Bailian Asset Holding. &萃苓 Community Shopping Center is the most critical one among the six project lines. Thanks to the advantages including the superior location and dense surrounding residential areas, No. 624 North Baoxing Road is positioned as a new dynamic community-based shopping center, which will also be the first flagship project of &萃苓. The renovation vision of "15-minute community life circle" is like a key to a better life, embracing the door of future dynamically and elegantly.

Dynamic Space

How to replace the single production function of this factory building with a new life function is the primary challenge for upgrading No. 624 North Baoxing Road. Reestablishing a life order matching with the surrounding area on this post-industrial site to attract people is the community business that Bailian Asset Holding is striving for. These old factories of different sizes are like a group of magic cube boxes, which will be endowed with new life functions after magical renovation.

Bailian Asset Holding has a clear understanding of the community business. First, community business is a local business which takes the residents in the community as the object of services, and facilitating, benefiting, and satisfying and promoting the comprehensive consumption of residents as the goal. Secondly, the community business will focus on providing one-stop daily life support and demand services for target consumers. Through differentiated and scenario-based experience, it will create an interesting and diversified life paradise and social platform for surrounding residents, and consider the consumption needs of the surrounding business customers to deliver more diversified business functions as a new community commercial landmark.

Bailian Asset Holding names its community commercial project lineup as 辛荟, a community communication space linking people, and an ecological three-dimensional block friendly to parents, children and pets. Everyone here would greet each other with a greeting: Nice to meet you (Xinghui)!

改造设计模型方案　Renovation design model

活力空间

北宝兴路624号项目升级的首要挑战是如何把这个厂房从单一的生产功能置换为新的生活功能？在废弃厂区的后工业遗址现场重新建立起一种与周遭匹配的生活秩序，打开这片沉寂已久土地的时空大门，让周边的人群愿意走进来，停下来，享受在这里的每分每秒，这就是百联资控想要做成的社区商业。这几幢规模大小不一的老厂房就像一群魔方盒子，将通过"魔力改造"后装入新的生活功能。

对于社区商业，百联资控有着清晰的认识。首先，它是以社区范围内的居民为服务对象，以便民、利民、满足和促进居民综合消费为目标的属地型商业。其次，社区型商业将侧重为目标消费客群提供一站式日常生活配套与需求服务，通过差异化和场景体验，为周边社区居民打造一个有趣、多元的生活乐园与社交平台，同时兼顾周边商务客群的消费所需，令社区商业功能更多元化，成为社区里的商业新地标。

百联资控为其社区商业产品线起了一个动人的名字——"莘荟"，一个链接着人与人的社区交流空间，一个亲子宠物友好的生态立体街区。在这里人人都会互相问候一句："幸会！"

改造后效果 renovation effect

唤醒 RENAISSANCE

The project site mainly covers Daning Road street, Liangcheng New Village street, New Gonghe Road street, and Guangzhong Road street, radiating more than 20,000 residents within a radius of 1.5km, coupled with over 10,000 teachers and students in the campus of Shanghai University on Yanchang Road. The population density is quite high. The Daning International Business Plaza, Daning Central Square and the upcoming Jiuguang Department Store are all positioned as large-scale regional comprehensive commerce, so residents living near the North Baoxing Road need to drive to arrive them before consumption. Every weekend and holiday, difficult parking, concentrated passenger flow, and increasingly expensive consumption spring up. A community neighborhood business center with walking distance, at a more friendly and pleasant scale, and featuring more frequent experience is still absent in this area—and therefore the 萃荟 project lineup exists.

Combined with the investigation of the actual living needs of the regional customers, Bailian Asset Holding has fostered a vision for No. 624 North Baoxing Road—building a public block open to the surrounding communities through necessary repair based on the original old factory building, in order to further open the original split community environment and optimize the urban interface. The 15-minute walking life circle will be taken as the scale of happiness to provide various supporting service facilities for residents in surrounding communities. The previously closed and passive factory can become a paradise for open shopping experience and parent-child life, and a community public space with a sense of belonging. Besides, through the linkage of the park, the business district, the community and the campus, a two-wheel drive of the section around Shanghai University and the business district around Daning provides high-level film and television cultural innovation and entrepreneurship talents with high standard hotel-style apartments and creative office space.

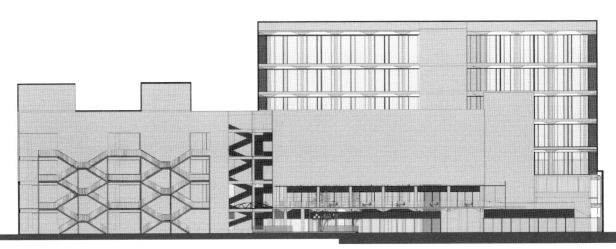

设计方案示意图　design ketch

项目所在地块主要辐射大宁路街道、凉城新村街道、共和新路街道、广中路街道，半径1.5公里范围内的住宅居民超过20000户，再加上上海大学延长路校区超过10000名师生人数，人口密度相当高。

尽管附近已有大宁国际广场、大宁中心广场以及待建成的久光百货，但均定位为大体量区域级综合商业，对北宝兴路周边居民来说，需要开车过去才能消费。每逢周末节假日，这些大型商业体就会出现停车难、客流集中等位、消费越来越贵等"大mall病"。一种步行距离可达、尺度上更亲切愉悦、体验上更高频的社区邻里商业中心在这片区域依旧缺位——而这，就是 莘苍 产品线想要做的。

结合区域客群实际生活需求的调研，百联资控为北宝兴路624号老厂房制订了一个愿景——以原有老厂房为基底，通过必要的修缮，打造一个向周边社区开放的公共街区，将原有割裂的社区环境通过进一步的开放，优化城市界面。以步行15分钟可达的生活圈为幸福感的尺度，提供周边社区居民生活所需的各项配套服务设施。让原先封闭消极的厂区成为提供开放式购物体验和家庭亲子生活的乐园，一个有归属感的社区公共空间。同时，以园区、商区、社区、校区四区联动，环上大板块和环大宁商圈双轮驱动，为高层次影视文化创新创业人才提供高标准的酒店式公寓和创意办公场所。

入口商业广场效果图　design sketch of the commercial plaza at the entrance

创新即新

如何将一个年轻而充满活力的商业场所植入庞大而封闭的冷库厂房？

百联资控始终围绕着"打造创新实体的社区体验式商业、满足社区需求的公园式邻里街区、多样化的空间形式赋予都市社交新体验"这些改造目标展开研究，细致梳理了项目改造的三大策略。

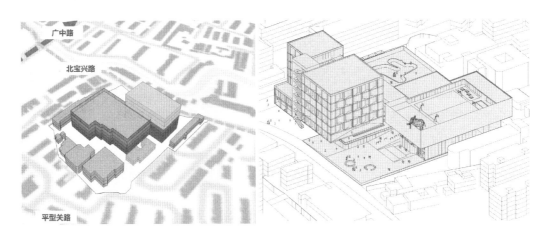

设计方案示意图 design ketch　　　　　　项目规划方案 project planning scheme

策略一 开放：激活与城市生活的联系

项目既有的建筑与工业遗存丧失了它原本的生产功能，想要激活它，就必须将当代公共空间和生活重新引入到已经丧失功能的旧建筑中，使之重新具备活力和价值。

策略二 介入：重塑建筑的空间架构

既有建筑改造通常需要面对历经岁月模糊而复杂的建筑状况，在设计中应发掘和提炼建筑的空间潜力和标识性，通过简练而有力的手段将空间阐明，给建筑一个空间结构和认识上的"骨骼"，使之真正成为当代生活所用的空间。

策略三 利用：挖掘建筑的空间潜力

老厂房通常具有有别于日常建筑的尺度和形制，对于这些特点的有效利用既是一种集约空间资源的手段，又可以增强空间的识别性和特征。

Innovative Renovation

How to transform a large closed cold storage into a young and dynamic business site?
Based on the renovation objective of "creating an innovative physical community experience-based commerce, a park-style neighborhood block that meets the community demand, and diversified space forms to give new social experience to Shanghai", Bailian Asset Holding has carefully presented three strategies for renovation.

Strategy 1 Opening: activate the connection with urban life
The existing buildings and industrial remains of this project have lost the original production function. To activate it, it is necessary to reintroduce the contemporary public space and life into this old building and endow it with new vitality and value.

Strategy 2 Intervention: reshaping the spatial structure of the building
Renovation usually needs to face the fuzzy and complex situation after years. In the design, it is necessary to explore and refine its space potential and identity, and clarify the space through simple and powerful means, aiming to design a framework in terms of spatial structure and understanding, making it a space used in contemporary life.

Strategy 3 Utilization: explore the space potential of the building
Old factories usually have different scales and shapes from routine buildings. The effective use of these characteristics is not only a means to intensify space resources, but also can enhance the identification and characteristics of space.

🐝 莘苔 内街效果图 design sketch of interior street in 🐝 莘苔

在具体的空间改造上，我们围绕"开放、连通、立体空间"三个要素做"加减法"。

"开放"，即打破外墙，将原来封闭的仓库内场向城市开放作为社区的活动广场，将场地内各种类型的室外堆场改造为引人驻足的活动场地。

"连通"，就是要创造廊道空间，将场地内各室外空间相连，形成可漫步、可逗留的开放式步行网络。

"立体空间"，则是将主体建筑的二层作为开放性邻里集市，与地面层的步行空间构成互动、立体的邻里中心。对建筑体量局部进行拆除，通过重组场地，创造开放空间，在低层形成穿越式动线，提供穿越式街区体验。

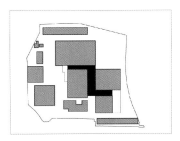

空间改造策略：建筑体量局部拆除
spatial renovation strategy: demolishing part of the building volume

空间改造策略：创造开放空间
spatial renovation strategy: creating open space

空间改造策略：提供穿越式街区体验
spatial renovation strategy: providing crossing block experience

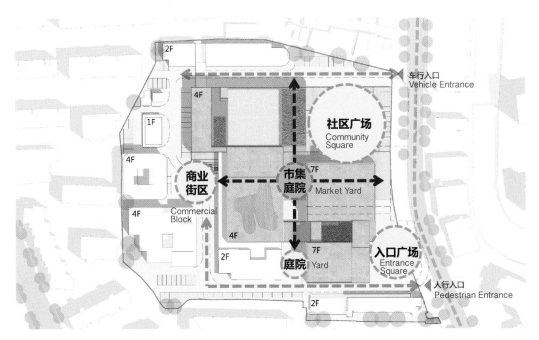

开放空间设计示意图
design sketch of the open space

In specific spatial renovation, "addition and subtraction" will be made based on three elements of "opening, connection and three-dimensional space".

"Opening" is to break the outer wall to open the original enclosed warehouse to the city as a community activity square, transforming various types of outdoor storage yards in the site into an attracting activity venue.

"Connection" is to create corridor space and connect the outdoor space in the site to form an open walking network that can walk and stay.

"Three-dimensional space" is to take the second floor of the main building as an open neighborhood market, forming an interactive and three-dimensional neighborhood center with the pedestrian space on the ground floor. Part of the building volume will be demolished to create open space by reorganizing the site, and form a crossing moving line at the ground floor for crossing block experience.

内庭院效果图
design sketch of interior yard

在功能配置上，原冷库与加工车间在一、二层形成开放的集市街区，同时置入工作坊，加工车间的三、四层为相对完整的集中商业体量，冷库主要作为办公使用，而外围的几栋零散建筑则将改造成长租公寓使用。

In functional configuration, the original cold storage and processing workshop will form an open market block on the first and second floors, together with the complete centralized commercial volumes on the third and fourth floors. The cold storage will be mainly for office use, whose peripheral scattered buildings will be transformed into long-term rental apartments.

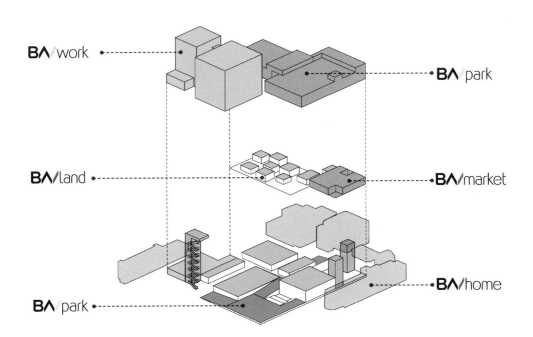

改造业态分布　distribution of renovated business forms

过去，老厂房只需要满足生产功能，如果增加商业功能，对环境舒适度的要求必然增高。设计师结合建筑的结构体系，有机拆分核心大体量，同时形成了几个各具特色的中庭或院落，引入大自然采光和景象，让空间更具灵动生机。

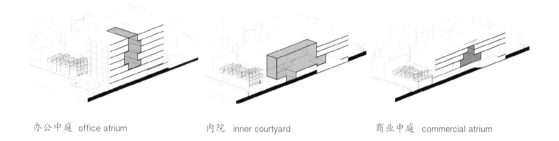

办公中庭 office atrium　　　　内院 inner courtyard　　　　商业中庭 commercial atrium

设计师将原有冷库穿堂部分进行局部拆除，扩大两侧的采光面，形成办公中庭；将原有冷库与加工车间之间的框架结构拆除，使得办公与商业之间形成开放的内院；还打断两个加工车间之间的联系，形成商业中庭，在具有交通功能的同时提供一定的开放公共空间。

The old factory used to only have the production function. If the business function is added, the requirement for environmental comfort will increase. Combined with the structural system of the building, the designer organically splits the core large volume into several atriums or courtyards with different characteristics. The introduction of natural lighting and scene makes the space more vivid.

The designer partially demolishes the original cold storage hall, and expands the lighting space on both sides to form the office atrium; demolishes the frame structure between the original cold storage and the processing workshop to form an open inner courtyard; interrupts the connection between the two processing workshops to form the commercial atrium, providing certain open public space in addition to the traffic function.

内街与庭院设计效果图
design sketch of interior street and yard

外立面的改造：在满足功能的情况下尽可能保留原有立面，采用黑色钢板、金属网、涂料、超白玻璃材料，既不抹去原有厂房的工业风貌和场所记忆，又以最集约高效的方式打造标识性沿街立面。

办公业态的建筑：利用冷库本身双层墙特性，通过开洞形式简化外立面改造，立面采用最常见的砌体墙与铝合金窗构造，施工简便，成本可控。商业区域的外立面则采用金属网，以简单的方式统一整个商业立面。

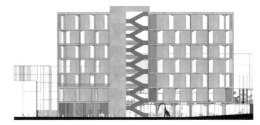

办公产品外立面设计方案图
design sketch of appearance of office building project

Renovation of the facade: The original appearance will be maximally kept on the premise of meeting the function; black steel plate, metal mesh, painting, and ultra-white glass materials will be used to create into a landmark facade along the road most intensively and efficiently while not sacrificing the industrial style and place memory of the original building.

Construction of office form: Using the characteristics of double-layer walls of the cold storage, the facade will be simplified in an opening form. The facade will be renovated with the most common masonry wall and aluminum alloy windows, featuring simple construction and controllable cost. The facade of the commercial area will adopt metal mesh to unify the style in a simple way.

莘荟 全景鸟瞰效果图
panoramic aerial view of 莘荟

商业产品外立面效果图
design sketch of the facade of the commercial project

品质生活

消费升级是中国目前最为明确的一大趋势，新生代人群对消费内容提出了更高要求，原有的物质消费早已让位给精神消费、文化消费和休闲消费。城市更新作为未来城市发展的重要内容，势必也要适应消费升级的趋势。

相比代表城市形象的购物中心，家门口的社区商业被认为是人们更高频消费的地方。我们生活在高层住宅里，可能是少有时间和机会去认识门对门的邻居，可能楼下便利店的店员比小区邻居更熟悉。而大部分上班族在工作日时间里，也会在以家为中心的固定半径里生活和消费。

这时候，社区商业就是一个创造具体、真实、可感受的幸福感的精致场所。在电商如此发达的今天，社区商业变成了受网购冲击最小的领域。它本就在场景打造中具备天然的情感优势，比城市商业更加注重体验感提升。

在消费升级的新零售时代，社区商业"必须是"周边生活人群的生活管家、安全员，提供实用性强并具有保障的生活服务。同时，"更应是"一个精致的会客厅、一个提升幸福感的课堂，成为提升生活品质的第一消费场所。未来，社区商业势必将成为"新生活运动"的引领者，为周边居民提供更多的活动空间，享受体验式邻里中心带来的新型社交体验。

莘荟从区域空白点突破，扮演人们生活方式提案者的角色，通过场景化的精心设计、业态品类品牌的科学筛选，补足区域欠缺资源，打造出一个精致温馨，小而美的社区邻里商业中心。相信它的出现将不仅改变现有社区商业的建筑形态和品牌矩阵，更深入到运营中，从而将社区商业推上新的阶梯，打造社区商业新生态，向"生活方式中心"转变。

Quality Life

Consumption upgrading is the clearest trend in China. Young people are putting forward higher requirements for consumption content. The material consumption has given way to spiritual consumption, cultural consumption and leisure consumption. As an important part of future urban development, urban renewal is bound to adapt to the trend of consumption upgrading.

Compared with the shopping center which represents the image of a city, the community business is considered to be consumed more frequently. We live in a high-rise house. We may not have enough time and opportunity to meet our door-to-door neighbors. Maybe the clerk in the convenience store downstairs is more familiar than the neighbors. Most office workers will also live and consume within a fixed radius centered on their house during working days.

Hence, community business is an exquisite place to create concrete, real and perceptible happiness. Today, with the development of e-commerce, the community business is least affected by online shopping for its natural emotional advantages in scenario building, and focus more on experience improvement than urban commerce.

In the new retail era of consumption upgrading, community business should be a housekeeper and security officer of the surrounding residents to provide practical and guaranteed life services. It more should be a delicate reception hall, a classroom that can enhance happiness, and an optimal consumption place to improve the quality of life. In the future, community business will surely become a leader of the "new life movement", providing more activity space for the surrounding residents and enjoying the new social experience brought by the experience-based neighborhood center.

Playing the role of lifestyle proponent from regional blank areas, 83萃荟 is dedicated to creating a small beautiful and warm community business center by making up for the lack of resources through scientific screening of the scenario-based careful design and business form brands. 83萃荟 is expected to not only change the building form and brand matrix of the existing community business, but go deeper into the operation, so as to push the community business to a new level, create a new community business ecology, and transform it into a "lifestyle center".

北宝兴路624号项目改造示意图
renovation sketch of No. 624 North Baoxing Road project

唤醒
RENAISSANCE

项目手记

　　试想一下这样的生活场景：阳光明媚的早晨，社区居民们可以在广场上晨练热身，开启活力的一天；中午可以在美食街寻觅最爱的那一口；下了班在生鲜超市里从容地挑选新鲜食材带回家烹煮一顿家庭大餐；周末，带着孩子和宠物来这里闲逛，或约上闺蜜好友小聚下午茶、享受一场放松身心的SPA；即便是深夜下班回家的路上，这里的24小时便利店里也有暖胃暖心的关东煮和热饮；下雨天也不必为了3公里以外的外卖等上半天……

　　原本家门口的废弃工厂变成了品质生活的舞台，曾经的机器轰鸣声切换成了如今悦耳动人的商场背景音乐，曾经一代人辛勤奋斗的地方成了活力四射、热情洋溢的街区，从"老工厂"到"生活场"的蜕变，正是城市更新的魅力和价值所在。时空转换，沉寂的老建筑又"复苏"过来，开始书写新生活的诗篇。

Project Notes

Imagine such a life scene: On a sunny morning, the community residents can warm up in the square to start a day with energy; at noon, they can eat something they like on the food street; after work, they can buy fresh ingredients in the supermarket calmly and cook them at home; at weekends, they can take their children and pets to hang out here, or meet their best friends and enjoy afternoon tea or a SPA; even on the way home from work late at night, the 24-hour convenience store here also has hot oden and drinks; in rainy days, they do not have to wait for a long time for order delivery 3 kilometers away...

The displaced factory has become a stage for quality life. The roar of machines has been turned into the background music of today's beautiful shopping malls. The place where a generation of people worked hard has become a vibrant and passionate block. The transformation from the "old factory" to the "living field" is exactly the charm and value of urban renewal. Time and space are changing, and the old silent building has revived to start a new life.

第八章
CHAPTER EIGHT

改造中的北宝兴路624号
No. 624 North Baoxing Road under renovation

【后记】
Postscript

后记 POSTSCRIPT

上海是一本打开的书

凭建筑便可读懂它独有的抱负

世界范围内的注视与对标

更赋予其诗意的场所精神、不可再生的美学价值！

感谢总有一些人

在心心念念着建筑与城市的双向唤醒

感谢总有一些人

在孜孜以求着建筑艺术与公共文化的极致耦合！

Shanghai is an opened book.

Its unique ambition can be read from buildings.

World-wide gazing and benchmarking

endows Shanghai with a poetic spirit of place and non-renewable aesthetic value!

We appreciate those

who are committed to the renaissance of both buildings and the city.

We appreciate those

who are always seeking after the ultimate coupling between architectural art and public culture!

唤醒
RENAISSANCE

上海，随着城市经济、产业升级的飞速发展，在新增土地供应量逐步紧缩、政府要求实现供地"零增长"甚至"负增长"的背景下，城市更新必将成为未来房地产发展的核心命题。开启地产新黄金时代的使命，离不开大量企业的共同推动和深度参与。

作为大型国企，百联集团多年来积极履行使命、承担社会责任，不仅见证了上海的发展与变迁，也在参与城市建设的过程中，一步一个脚印，步步抢占先机；看似机缘巧合，实则提前布局积累了丰富的城市更新经验，成功推进了一大批精品项目，成为上海城市更新生力军中当仁不让的典范。

百联资控的城市更新之路，也随着上海城市建设的进度，从量变到质变，从演变到嬗变的升华过程，迎接着难得的机遇和巨大的挑战。城市更新政策的变化频率和细分程度日益陡增，如何才能紧跟形势、融入政府发展战略，抓住历史性机遇，彰显国企的社会责任，伴随着国家和区域的腾飞，实现共同发展，这一切，无疑对我们在全球视野、运营能力、综合素质等诸多方面，提出了更高的要求。

城市更新，是城市的又一次成长，也是企业的又一次成长。从生长、繁荣、更新到再生，为了人民对美好生活的向往，为了重塑都市活力，焕发都市新生，我们将继续不畏艰难、奋勇争先，为上海城市更新奉献自己的一份力量，赋予这片魔力热土更多美好梦想！

With the rapid development of urban economy and industrial upgrading, Shanghai is facing the restrictive supply of new land and the government's request for "zero growth" or even "negative growth" of land supply. Urban renewal is bound to become the core proposition of future real estate development. The mission of entering the new golden age of real estate industry cannot be without joint efforts and wholehearted participation of a large number of enterprises.

As a large state-owned enterprise, Bailian Group has always vigorously fulfilled its mission and assumed social responsibility. It has not only witnessed the development and evolution of Shanghai, but also participated in the process of urban construction. Working steadily, Bailian Group has seized the first-mover opportunity by steps to draw a blueprint in advance under seeming coincidence. Having gained rich experience in urban renewal and successfully launched many quality projects, Bailian Group has grown into a model among new comers in this industry.

With the progress of urban construction, the urban renewal road of Bailian Group has sublimed from quantitative change to qualitative change, and from evolution to metamorphosis, echoing with the rare opportunities and great challenges. The changing frequency and segmentation of urban renewal policies are increasingly sharp. How to keep abreast of the situations and respond to the governmental development strategy, seize the historic opportunity, show the social responsibility of a state-owned enterprise, go after the rapid local and national development, and achieve common development? All of these undoubtedly have put forward higher requirements for Bailian Group in many aspects, such as global vision, operational capability and comprehensive quality.

Urban renewal is new growth of the city and enterprises. From growth, prosperity, renewal to rebirth, in order to respond to people's aspiration for a better life, to revive the vitality of Shanghai and rejuvenate the city, we will continue to contribute to the urban renewal of Shanghai through thick and thin, and help this magical place realize its more beautiful dreams!

参考文献:
References

[1] 郑时龄.上海的城市更新与历史建筑保护[J].中国科学院院刊,2017,32(7):690-695.
[2] 郑时龄.上海近代建筑风格[M].上海:上海教育出版社,1995.
[3] 蔡育天.回眸[M].上海:上海人民出版社,2001.
[4] 承载,吴健熙.老上海百业指南——道路机构厂商住宅分布图[M].上海:上海社会科学院出版社,2008.
[5] 朱国栋,刘红.百年沪商[M].上海:上海财经大学出版社,2010.
[6] 杨嘉祐.上海老房子的故事[M].上海:上海人民出版社,1999.
[7] 沈福煦,沈燮癸.透视上海近代建筑[M].上海:上海古籍出版社,2004.
[8] 王垂芳.洋商史:上海1843-1956[M].上海:上海社会科学院出版社,2007.
[9] 叶又红.海上旧闻:第2辑[M].上海:文汇出版社,2000.
[10] Edward Denison, Guang Yu Ren.Building Shanghai:The Story of China's Gateway[M].London:Wiley Academy,2006.
[11] 汪耀华.留存着的书业时光[M].上海:上海书店出版社,2016.
[12] 惜珍.永不拓宽的上海马路(一)[M].上海:东方出版中心,2016.
[13] 庄智娟.湖社和陈英士纪念堂[J].档案春秋,2009(2):56-57.
[14] 薛顺生.杨树浦路上老仓库——原马登仓库与"61号老栈房"[J].都会遗踪,2010(1):37-39.
[15] 上海市人民政府.上海市城市总体规划(2017-2035年)[R/OL].(2018-01-04)
     [2019-12-16].http://www.shanghai.gov.cn/nw2/nw2314/nw32419/nw42806/.
[16] 城市百网科.老上海的钱庄街[EB/OL].http://www.csbkw.com/shanghai/lishi/gushi/18605.html.
[17] 新华报业网.郑钧天."上海金融科技园区"揭牌将构建金融科技生态圈[EB/OL].
     (2019-03-24)[2019-12-16].http://news.xhby.net/system/2019/03/24/030938998.shtml.

图书在版编目（CIP）数据

唤醒：穿越历史 邂逅经典 / 程大利主编.
—上海：同济大学出版社，2019.12
ISBN 978-7-5608-8867-5

Ⅰ.①唤… Ⅱ.①程… Ⅲ.①城市史－建筑史－上海
Ⅳ.①TU-098.12

中国版本图书馆CIP数据核字(2019)第262181号

唤醒——穿越历史 邂逅经典（百联资控城市更新实录）

主　编	程大利
执　笔	程大利　刘懿　李再励
出　品	百联资产控股
策　划	Insight见著
顾　问	施文球
指　导	曾浙一
采　编	徐陆薇
编委会	程大利　施文球　刘懿　倪群　王刚　杨耀宇　徐陆薇　李再励 徐其明　富永直树　卞泽洪　朱蒙　邢宪斌
特别鸣谢	上海市历史建筑保护事务中心
责任编辑	吕炜　马继兰
责任校对	徐春莲
封面设计	杨耀宇
装帧设计	Insight见著
图片摄影	常念祖　邢振中　殷菲
插图设计	陈伶鹤
英文翻译	北京博思译言信息技术有限公司

出版发行	同济大学出版社　www.tongjipress.com.cn （上海市四平路1239号　邮编：200092　电话：021-65985622）
经　销	全国各地新华书店
印　刷	上海安枫印务有限公司
开　本	889mm×1194mm　1/16
印　张	16.75
字　数	536000
版　次	2019年12月第1版　2019年12月第1版印刷
书　号	ISBN 978-7-5608-8867-5
定　价	198.00元

版权所有　侵权必究　印装问题　负责调换